Mining Human Mobility in Location-Based Social Networks

Synthesis Lectures on Data Mining and Knowleddte Discovery

Editors
Jiawei Han, *University of Illinois at Urbana-Champaign*
Lise Getoor, *University of Maryland*
Wei Wang, *University of North Carolina, Chapel Hill*
Johannes Gehrke, *Cornell University*
Robert Grossman, *University of Chicago*

Synthesis Lectures on Data Mining and Knowledge Discovery is edited by Jiawei Han, Lise Getoor, Wei Wang, Johannes Gehrke, and Robert Grossman. The series publishes 50- to 150-page publications on topics pertaining to data mining, web mining, text mining, and knowledge discovery, including tutorials and case studies. Potential topics include: data mining algorithms, innovative data mining applications, data mining systems, mining text, web and semi-structured data, high performance and parallel/distributed data mining, data mining standards, data mining and knowledge discovery framework and process, data mining foundations, mining data streams and sensor data, mining multi-media data, mining social networks and graph data, mining spatial and temporal data, pre-processing and post-processing in data mining, robust and scalable statistical methods, security, privacy, and adversarial data mining, visual data mining, visual analytics, and data visualization.

Mining Human Mobility in Location-Based Social Networks
Huiji Gao and Huan Liu

ISBN: 978-3-031-00780-4   paperback
ISBN: 978-3-031-01908-1   ebook

DOI 10.1007/978-3-031-01908-1

A Publication in the Springer series
SYNTHESIS LECTURES ON DATA MINING AND KNOWLEDGTE DISCOVERY

Lecture #11
Series Editors: Jiawei Han, *University of Illinois at Urbana-Champaign*
              Lise Getoor, *University of Maryland*
              Wei Wang, *University of North Carolina, Chapel Hill*
              Johannes Gehrke, *Cornell University*
              Robert Grossman, *University of Chicago*
Series ISSN
Print 2151-0067   Electronic 2151-0075

Mining Human Mobility in Location-Based Social Networks

Huiji Gao
LinkedIn

Huan Liu
Arizona State University

SYNTHESIS LECTURES ON DATA MINING AND KNOWLEDGTE DISCOVERY #11

ABSTRACT

In recent years, there has been a rapid growth of location-based social networking services, such as Foursquare and Facebook Places, which have attracted an increasing number of users and greatly enriched their urban experience. Typical location-based social networking sites allow a user to "check in" at a real-world POI (point of interest, e.g., a hotel, restaurant, theater, etc.), leave tips toward the POI, and share the check-in with their online friends. The check-in action bridges the gap between real world and online social networks, resulting in a new type of social networks, namely location-based social networks (LBSNs). Compared to traditional GPS data, location-based social networks data contains unique properties with abundant heterogeneous information to reveal human mobility, i.e., "when and where a user (who) has been to for what," corresponding to an unprecedented opportunity to better understand human mobility from spatial, temporal, social, and content aspects. The mining and understanding of human mobility can further lead to effective approaches to improve current location-based services from mobile marketing to recommender systems, providing users more convenient life experience than before. This book takes a data mining perspective to offer an overview of studying human mobility in location-based social networks and illuminate a wide range of related computational tasks. It introduces basic concepts, elaborates associated challenges, reviews state-of-the-art algorithms with illustrative examples and real-world LBSN datasets, and discusses effective evaluation methods in mining human mobility. In particular, we illustrate unique characteristics and research opportunities of LBSN data, present representative tasks of mining human mobility on location-based social networks, including capturing user mobility patterns to understand when and where a user commonly goes (location prediction), and exploiting user preferences and location profiles to investigate where and when a user wants to explore (location recommendation), along with studying a user's check-in activity in terms of why a user goes to a certain location.

KEYWORDS

location-based social networks, human mobility, social media, data mining, location recommendation, location prediction

This effort is dedicated to my parents and Ye.
Thank you for everything. –HG

To my parents, wife, and sons. –HL

Contents

Acknowledgments

First and foremost, we would like to acknowledge many colleagues at the Data Mining and Machine Learning Lab of Arizona State University for their helpful suggestions and contributions in various ways to this time-consuming book project. It is a great pleasure working with them, particularly, Jiliang Tang, Xia Hu, Lei Tang, Zheng Zhao, XufeiWang, Geoffery Barbier, Shamanth Kumar, Fred Morstatter, Ali Abbasi, Reza Zafarani, Pritam Gundecha, Isaac Jones, Suhas Ranganath, and Salem Alelyani.

We thank Morgan & Claypool and particularly executive editor Diane D. Cerra for her help and patience throughout this project. This work is part of the projects sponsored by grants from ONR (No. N000140810477, No. N000141010091, and No. N000141110527) and Graduate Education Dissertation Fellowship.

Last but not the least, we thank our families for their support through this entire project. We dedicate this book to them, with love.

Huiji Gao and Huan Liu
April 2015

Figure Credits

Figure 1.2	Based on: H. Gao and H. Liu. Data analysis on location-based social networks. In Mobile Social Networking, pages 165–194. Springer, 2014.
Figure 1.3	Based on: H. Gao, J. Tang, and H. Liu. gSCorr: Modeling geo-social correlations for new check-ins on location-based social networks. 21st ACM International Conference on Information and Knowledge Management, 2012.
Figure 1.4	From: A. Sadilek, H. Kautz, and J. Bigham. Finding your friends and following them to where you are. In Proceedings of the Fifth ACM International Conference on Web Search and Data Mining, pages 723–732. Copyright © 2012 ACM. Used with permission.
Figure 1.5	Based on: Z. Yin, L. Cao, J. Han, C. Zhai, and T. Huang. Geographical topic discovery and comparison. In Proceedings of the 20th International Conference on World Wide Web, pages 247–256. ACM, 2011.
Figure 1.6	Based on: T. Sakaki, M. Okazaki, and Y. Matsuo. Earthquake shakes twitter users: real-time event detection by social sensors. In Proceedings of the 19th International Conference on World Wide Web, pages 851–860. ACM, 2010.
Figure 1.7	From: S. Kumar, F. Morstatter, and H. Liu. Twitter Data Analytics. Copyright © 2013, Springer, New York, NY, USA, 2013. Used with permission.
Figure 1.8	From: A. Sadilek, H. Kautz, and V. Silenzio. Modeling spread of disease from social interactions. In Sixth AAAI International Conference on Weblogs and Social Media (ICWSM), 2012. Copyright © 2012 AAAI. Used with permission.
Figure 2.3	Based on: S. Scellato, A. Noulas, R. Lambiotte, and C. Mascolo. Socio-spatial properties of online location-based social networks. Proceeding of the 5th International AAAI Conference on Weblogs and Social Media, 11, 2011.

Figure 2.4	Based on: S. Scellato, A. Noulas, R. Lambiotte, and C. Mascolo. Socio-spatial properties of online location-based social networks. Proceeding of the 5th International AAAI Conference on Weblogs and Social Media, 11, 2011.
Figure 2.5	From: H. Gao, J. Tang, and H. Liu. Personalized location recommendation on location-based social networks. In Proceedings of the 8th ACM Conference on Recommender systems, pages 399–400. Copyright © 2014 ACM. Used with permission.
Figure 2.6	Based on: M. Ye, K. Janowicz, C. Mülligann, and W. Lee. What you are is when you are: the temporal dimension of feature types in location-based social networks. In Proceedings of the 19th ACM SIGSPATIAL International Conference on Advances in Geographic Information Systems, pages 102–111. ACM, 2011.
Figure 2.7	Based on: Z. Cheng, J. Caverlee, K. Lee, and D. Sui. Exploring millions of footprints in location sharing services. In Proceedings of the Fifth International Conference on Weblogs and Social Media, 2011.
Figure 2.8	Based on: L. Backstrom, E. Sun, and C. Marlow. Find me if you can: improving geographical prediction with social and spatial proximity. In Proceedings of the 19th International Conference on World Wide Web, pages 61–70. ACM, 2010.
Figure 2.9	Based on: Z. Cheng, J. Caverlee, K. Lee, and D. Sui. Exploring millions of footprints in location sharing services. In Proceedings of the Fifth International Conference on Weblogs and Social Media, 2011.
Figure 2.10	Based on: E. Cho, S. Myers, and J. Leskovec. Friendship and mobility: user movement in location-based social networks. In Proceedings of the 17th ACM SIGKDD International Conference on Knowledge Discovery and Data Mining, pages 1082–1090. ACM, 2011.
Figure 2.11	Based on: H. Gao, J. Tang, and H. Liu. Exploring social-historical ties on location-based social networks. In Proceedings of the Sixth International Conference on Weblogs and Social Media, 2012.
Figure 2.12	Based on: Z. Cheng, J. Caverlee, K. Lee, and D. Sui. Exploring millions of footprints in location sharing services. In Proceedings of the Fifth International Conference on Weblogs and Social Media, 2011.

Figure 2.13   Based on: Z. Cheng, J. Caverlee, K. Lee, and D. Sui. Exploring millions of footprints in location sharing services. In Proceedings of the Fifth International Conference on Weblogs and Social Media, 2011.

Figure 2.14   Based on: Z. Cheng, J. Caverlee, K. Lee, and D. Sui. Exploring millions of footprints in location sharing services. In Proceedings of the Fifth International Conference on Weblogs and Social Media, 2011.

Figure 2.15   From: E. Cho, S. Myers, and J. Leskovec. Friendship and mobility: user movement in location-based social networks. In Proceedings of the 17th ACM SIGKDD International Conference on Knowledge Discovery and Data Mining, pages 1082–1090. Copyright © 2011 ACM. Used with permission.

Figure 2.16   Based on: E. Cho, S. Myers, and J. Leskovec. Friendship and mobility: user movement in location-based social networks. In Proceedings of the 17th ACM SIGKDD International Conference on Knowledge Discovery and Data Mining, pages 1082–1090. ACM, 2011.

Figure 2.17   Based on: H. Gao, J. Tang, X. Hu, and H. Liu. Modeling temporal effects of human mobile behavior on location-based social networks. In Proceedings of the 22nd ACM International Conference on Information & Knowledge Management, pages 1673–1678. ACM, 2013.

Figure 2.18   Based on: H. Gao, J. Tang, X. Hu, and H. Liu. Modeling temporal effects of human mobile behavior on location-based social networks. In Proceedings of the 22nd ACM International Conference on Information & Knowledge Management, pages 1673–1678. ACM, 2013.

Figure 3.1   Based on: Stpasha. Normal distribution. http://en.wikipedia.org/wiki/Normaldistribution, 2008.

Figure 3.2   Based on: H. Gao, J. Tang, and H. Liu. Mobile location prediction in spatio-temporal context. Nokia Mobile Data Challenge Workshop, 2012.

Figure 3.3   Based on: H. Gao, J. Tang, and H. Liu. Mobile location prediction in spatio-temporal context. Nokia Mobile Data Challenge Workshop, 2012.

Figure 3.4   From: H. Gao, J. Tang, and H. Liu. Exploring social-historical ties on location-based social networks. In Proceedings of the Sixth International Conference on Weblogs and Social Media, 2012. Copyright © 2012 AAAI. Used with permission.

Figure 3.5   Based on: L. Backstrom, E. Sun, and C. Marlow. Find me if you can: improving geographical prediction with social and spatial proximity. In Proceedings of the 19th International Conference on World Wide Web, pages 61–70. ACM, 2010.

Figure 4.2   Based on: M. Ye, P. Yin, W. Lee, and D. Lee. Exploiting geographical influence for collaborative point-of-interest recommendation. In Annual International ACM SIGIR Conference on Research and Development in Information Retrieval, pages 325–334, 2011.

Figure 4.4   From: H. Gao, J. Tang, X. Hu, and H. Liu. Exploring temporal effects for location recommendation on location-based social networks. In Proceedings of the 7th ACM Conference on Recommender Systems, pages 93–100. Copyright © 2013 ACM. Used with permission.

Figure 4.5   Based on: H. Gao, J. Tang, and H. Liu. Personalized location recommendation on location-based social networks. In Proceedings of the 8th ACM Conference on Recommender systems, pages 399–400. ACM, 2014.

CHAPTER 1

Introduction

1.1 HUMAN MOBILITY BEHAVIOR

Human mobility, defined as "**when** and **where** a user (**who**) has been to for **what**," reflects the mobile aspect of human behavior in real world. Mining human mobility is affected by various factors. As the most significant factor in "mobility," location ("where") plays an important role in presenting humans' daily life and revealing their mobile preferences for studying human mobility.

1.1.1 LOCATION, LOCATION, LOCATION

Location, as the spatial characteristic of the world, has been considered as one of the most important factors for life and business. For many businesses, getting the right location can result in the difference between success and failure due to various location-based factors such as transportation, local culture, and natural resources. For example, a restaurant needs to consider the taste and dining habits of their customers, as well as consider the population around its location to estimate the number of potential customers. When choosing the location of a company, it is critical to consider its distance to local residents and communities which directly relates to its future, as most people are more willing to work locally than remotely and it is easier to recruit local workers.

When associated with humans, location is a key indicator of human mobility. The study of human mobility patterns on GPS data indicates that, on average, a person goes to dozens of places a week and hundreds of places a year [35]. Such frequent human interactions with geographical locations generate abundant information of human mobile preferences, suggesting great opportunities to study human mobility for designing advanced location-based services. For example, via capturing a user's previously visited restaurants at one location, one can recommend her new restaurants that she may be interested in for her future visit at another location. The user could benefit from the recommendation by reducing the exploring time with high possibility of enjoying nice food. The restaurant could also benefit from the increasing number of customers. Under the situation of disaster relief, by analyzing the normal-time human distribution at the disaster scene before a disaster, relief organizations can make fast responses on evacuation routes, and estimate the required resources (e.g., food, water, tents, medical assistances) of the affected area for effective disaster relief.

1.1.2   INFERRING HUMAN LIFESTYLES THROUGH LOCATIONS

Users visit different locations for vacations and entertainment. The large number of visited locations record a user's footprints. They make it possible to analyze her lifestyle and understand the "where" aspect of her mobility, such as the most frequently visited locations. Human mobility is commonly treated as a stochastic process. These footprints can introduce noise and complicate the extraction of mobility patterns. Thus, we need additional information to help better capture human mobility.

If temporal information (e.g., time-stamps of the user's visited locations) is available, one can study the "when" aspect of human mobility. For example, by chronologically connecting a user's visited locations during the previous week, a week-long location trajectory emerges. Such a location can be used to study like most frequent location sequences, e.g., shopping at a supermarket after having dinner at a restaurant. Another example is to study the periodic patterns of human mobility. By observing when a user visits locations, one can infer hourly patterns of a day, or daily patterns of a week, e.g., watching movie around 9 pm on Saturday. The above two examples correspond to the chronological and periodic mobility patterns embedded in the temporal information.

If social information (e.g., making friends, visiting a location with friends) is available, one could infer the co-visiting behavior, and study the "social" aspect of human mobility. Observing location-based content information, e.g., comments left at a restaurant, tips towards a shopping mall, can help us explore human mobility through a content view and understand the "what" aspect of human mobility regarding "what the user visits a location for."

1.1.3   MINING HUMAN MOBILITY WITH CELLPHONE DATA

The above examples exhibit a clear picture of mining human mobility from spatial, temporal, social, and content perspectives. Traditionally, such analysis is performed on cellphone-based GPS data. In the mobile era, cellphones have been widely used to facilitate communication and activities. A user's cell phone is often with him most of the time. Thus, cellphones are mobile sensors of human beings, while data collected through these sensors provides information regarding those "where," "when," "who," and "what" aspects. Such data of human mobility has led to location-based applications such as traffic forecasting, location prediction, and recommender systems. Typical cellphone-based GPS data contains a set of time-stamped GPS points that a user has been to, along with the mobile activities such as listening to music, generating Bluetooth connections, browsing web pages and watching videos. Since such data is obtained from users' cell phones through telecommunication services, user privacy is a big concern which limits the data availability. Users may not feel comfortable sharing their mobile data even for research purposes. Thus, work based on cellphone data commonly confronts the following limitations.

1. **Small-Scale Mobility Data**
   Cellphone-based GPS data generally contains a small number of users due to privacy concerns, which usually cannot be publicly available. The observations on such data may be

biased due to certain factors such as region, demography, gender, age, and education. For generating statistically significant conclusions especially in big data era, more data are encouraged when analyzing human mobility.

2. **Absence of Semantic Indications**
   GPS data store location information in terms of geographical coordinates, i.e., latitude and longitude. It is not straightforward to associate such coordinates with real-world points of interest, e.g., restaurants, hotels, theaters, and malls. Generally, semantic information of locations is not a available on GPS data. Thus, it is impractical to investigate the "what" aspect of human mobile behavior without such information. Although one can use third-party library to map coordinates into POIs, it does not work well on places with dense POIs, as it is difficult to distinguish POIs close to each other based on geographical coordinates. Furthermore, via observing that a user has stopped at a geographical point, it is not easy to determine whether he was visiting the corresponding POI or just passing by.

3. **Insufficient Social Information**
   Social connections are not easily obtained from GPS data. Generally, social connections can be inferred through the history of one's phone calls, messages, or bluetooth connections. However, it is difficult to collect this kind of data due to the user privacy concerns. There is work studying social connections on GPS data with small number of users participated who grant permissions to use their data. It is not encouraged to make such data available for other researchers to reuse. In addition, social information obtained through this way may be of low quality. For example, Bluetooth may not be commonly used thus connections inferred through this way may be biased; users who have phone communications do not necessarily indicate their friendships, not to mention that they share common interests (a general assumption of social theories used in social recommendation). On the other hand, social activities, such as friends visiting the same location together, are not available on GPS data either.

1.2   LOCATION-BASED SOCIAL NETWORKS

In recent years, location-based social networks (LBSNs) have attracted much attention from both academia and industry. LBSN data exhibits many advantages over the limitations of GPS data. Generally, a social network is a social structure consisting of nodes (e.g., individuals or organizations) and relationships (e.g., friendship or siblings) among these nodes. Social networks can be built through social networking services such as Facebook, Twitter, Linkedin, Google+, etc. LBSNs refer to one type of social networks with geographical properties embedded, which is usually generated through using location-based social networking services.

1.2.1   LOCATION-BASED SOCIAL NETWORKING SERVICES

Location-based social networking services, e.g., Foursquare,[1] Yelp,[2] and Facebook Places,[3] have emerged in recent years and become increasingly popular since 2010. The first commercial location-based social networking service available in the U.S. was Dodgeball,[4] launched in 2000. It allowed users to "check-in" by broadcasting their current locations through short messages to their friends who were within a 10-block radius; users could also send "shouts" to organize a meeting among friends at a specific place. Dodgeball was acquired by Google in 2005. In 2009, Google launched its location-based social networking services, "Google Latitude," while the founder of Dodgeball launched a new location-based social networking service "Foursquare" in the same year. Foursquare utilizes a game mechanism in which users can compete for virtual positions, such as mayor of a city, based on their check-in activities. It had reached 50 million users as of May, 2014,[5] becoming one of the most successful location-based social networking sites in the United States. Facebook also launched its location-based service, Facebook Places, in 2010 and acquired another popular LBSN service, Gowalla,[6] at the end of 2011.

Location-based social networking services maintain a large POI database and allow a user to "check-in" at a POI with his smartphone regarding to his current physical location. The user can also leave tips and share "check-in" experience with his online friends, along with creating the opportunity to make new friends. As reported by the Pew Internet and American Life Project, smartphone ownership among American adults rose from 35% in 2011 to 46% in 2012, while 18% of smartphone owners used location-based social networking services [120]. It is anticipated that location-based marketing in North America will grow from $1.8 billion in 2013 to $3.8 billion in 2018 [66]. Such a rapid growth of location-based social networking services has provided abundant information and greatly enriched the availability of human mobility data, making it possible to study human mobility on a large scale.

The "check-in," recognized as a location-tagging function, has been considered a distinguishing feature between location-based social networks and general social networks (GSNs). However, the boundary between these two types of social networks recently becomes blurry. By incorporating the location-tagging function into general social networks, it could be ambiguous to differentiate a LBSN from a GSN. For example, Twitter, as a typical general social network, launched Twitter Places which allows a user to tag their locations when tweeting as well as viewing tweets from a particular location [51]. Yelp enabled check-ins in 2010 and became a new LBSN after Foursquare and Gowalla [85]. In the future, with the rapid development of mobile technologies, it can be expected that LBSN may eventually be integrated into a location-based feature of a general social network.

[1] http://foursquare.com/
[2] http://www.yelp.com/
[3] http://www.facebook.com/about/location/
[4] http://en.wikipedia.org/wiki/Dodgeball
[5] https://foursquare.com/about
[6] http://en.wikipedia.org/wiki/Gowalla

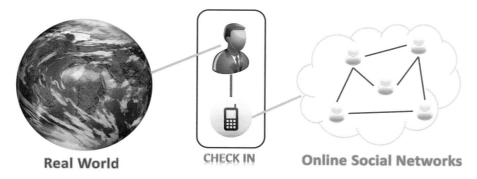

Real World   **CHECK IN**   **Online Social Networks**

Figure 1.1: Check-in actions connect the real world and a virtual world.

1.2.2   REAL WORLD VS. VIRTUAL WORLD

Typical location-based social networking services leverage Web 2.0 technology and mobile devices to facilitate a user's daily life when exploring the city and neighborhood. They provide various activities for a user to perform, such as "checking-in" at current locations, seeking local points of interest and discounts, leaving comments on specific places, making friends, and sharing check-in experience online. The "check-in" reports a user's visit at a physical place, and publishes such visits as an online post to let his friends know.

"Check-in" actions bridge the real world and the virtual world. Compared with many other online activities (following, liking, grouping, voting, tagging, etc.) that interact with the virtual world, "check-in" reflects a user's geographical action in the real world, residing where the online world and real world intersect, as shown in Figure 1.1. Thus, data collected from location-based social networks provides an ideal environment to analyze users' real world behavior through virtual media, which could potentially improve location-based services such as mobile marketing [5, 6, 84], disaster relief [23, 32, 36, 115], traffic forecasting [7, 20], and location recommendation [31, 105, 117].

1.2.3 W^4 INFORMATION LAYOUT

Generally, LBSNs consist of social connections among users and "location-based" context including geographical check-in POIs, check-in time stamps, and user/location-associated content (e.g., descriptions, tips, pictures, videos, etc.). Such elements reflect aspects of human mobility in a "W^4" (i.e., **who, when, where**, and **what**) information layout, corresponding to four distinct information layers, as shown in Figure 1.2.

The geographical layer contains the historical geographical locations of users, the social layer contains social friendships among users, the temporal layer indicates the time stamps of the users' "check-in" behavior, and the content layer consists of user-generated content such as tips

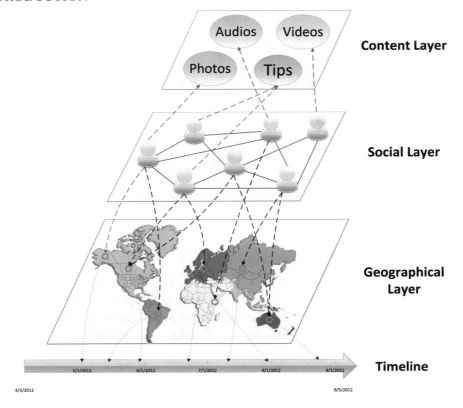

Figure 1.2: The information layout of LBSNs (based on ([24])).

about different places, photos taken during check-in, and location descriptions. LBSN data has an additional geographical layer which is not available in traditional online social networks [44], and an explicit social layer which is not available from mobile phone data [12]. The unique geographical property and the social network information presents new challenges for mining human mobility on LBSN data, since traditional approaches on social network or mobile phone data may fail due to the lack of pertinence.

The "W^4" defines six different types of networks, i.e., location-location, user-user, content-content (e.g., word-word), user-location, user-content, and location-content networks. Each network can be associated with the temporal layer, indicating more opportunities and challenges for mining human mobility on LBSNs from spatial, temporal, social, and content perspectives. Therefore, data mining techniques specifically designed for LBSNs can efficiently deal with these distinct properties, and help understand human mobility for research and application purposes.

1.3   COMPUTATIONAL TASKS

In this section, we discuss research and application opportunities on LBSNs, and introduce two fundamental computational tasks for mining human mobility with LBSN data.

1.3.1   RESEARCH AND APPLICATION OPPORTUNITIES

The rapid development of LBSNs in recent years provides researchers with opportunities to investigate human mobility from LBSN data w.r.t. various perspectives including spatial, temporal, social, and content. Many mobile applications have been developed to explore humans' mobile interests and facilitate their daily life. In this section, we introduce a set of research and application topics on LBSNs.

Recommender Systems

They are designed to recommend items to users in various scenarios such as watching movies, shopping items, exploring restaurants, and attending social events such as dating. Since the development of city and neighborhood offers more choices of life experience than before, recommendation is indispensable in helping users filter uninterested items and save time in decision making. Recommender systems on location-based social networks have become popular in recent years, while location (or POI) recommendation and friend recommendation are the major recommendation tasks.

In location recommendation, locations are recommended to a user according to his mobile interests. Contexts can be considered such as the current time and geographical position of the user, the user's previous check-ins, and the user's social friends' check-ins [31, 91]. Generally, the recommended locations in this task are new locations. Figure 1.3 illustrates a location recommendation scenario. A user u went to location L_1, L_2, and L_3 at time t_1, t_2, and t_3, the location recommendation system recommends a new location $L_4 \notin \{L_1, L_2, L_3\}$ to the user at time t_4. More will be discussed in Chapter 4.

Friend recommendation analyzes the common interests between users under the assumption that friends share more interests than non-friends. Such a property is derived from homophily [69], a social theory indicating that users tend to associate with others who behave in similar ways. Figure 1.4 illustrates a scenario of friend recommendation. Users are represented by pins on the map and the red links denote social connections. Each user is associated with geographical information (e.g., current position), behavior information (e.g., check-in history), and network information (e.g., existing friendships), these types of information can be leveraged to study the user's personal preferences for friend recommendation.

Geographical Topic Analysis

It investigates the semantic meaning of geographical regions, in order to enrich the functional description of locations for designing advanced location-based services. The diversity of geographical regions can be studied through the different distributions of location-based topics over areas. For

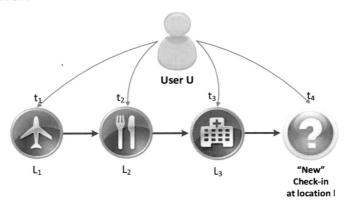

Figure 1.3: Location recommendation on LBSNs (based on ([29])). A user went to location L_1, L_2, and L_3 at time t_1, t_2, and t_3, the location recommendation system aims to recommend a new location to him at t_4.

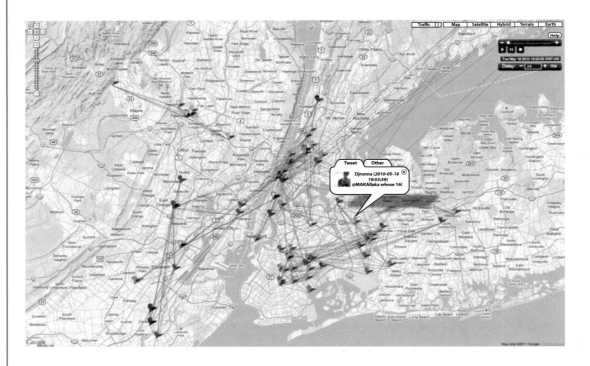

Figure 1.4: A sample of geo-active friends in New York City (based on ([78])). The links between users represent friendships.

example, coast, dessert, and mountain are the three typical topics which cover different areas in the U.S., as shown in Figure 1.5. Researchers utilize computational models to explore the spatial-temporal patterns of topical content [41, 75, 107]. Online systems, such as Livehoods,[7] have also been developed to explore the social dynamics of the city and reveal the different characterized regions [19]. In Livehoods, a clustering approach is adopted to cluster check-in locations from millions of check-ins into different areas, with each area representing a lifestyle character in the corresponding area.

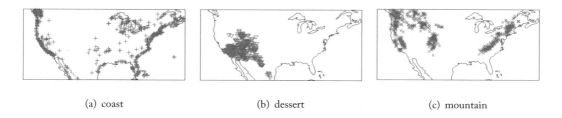

(a) coast   (b) dessert   (c) mountain

Figure 1.5: Three typical geographical topics (coast, dessert, and mountain) in the U.S. (based on ([107])).

Event Detection

It automatically detects events from large-scale and sparse datasets. For example, disaster event detection is one of the widely studied topics. During a disaster like earthquake or tsunami, messages posted from the disaster scene on LBSNs can be used to predict the disaster impacts, such as duration, degree, and coverage. Various systems have been proposed to detect and report disasters. For instance, Toretter [80] is an earthquake reporting system in Japan. It considers each user who tweets about a target event as a sensor of the event, and tracks the event center and trajectory with a spatio-temporal model. Figure 1.6 presents an estimation of affected areas during Japan earthquake in August 2009. TweetTracker[8] is an online tool designed to detect, track, analyze, visualize, and understand events on Twitter. It organizes tweets into events to facilitate near real-time tweet aggregation and to support search and analysis of the collected tweets. The system also allows a variety of visualizations based on this information, including streaming geo-spatial visualization, word cloud summarizations, post-event investigations in pseudo real-time, automatic translation of non-English tweets, and keyword trending and comparison. Figure 1.7 shows an illustrative screenshot of TweetTracker. Disease spread detection systems have also been developed to capture the spread of disease [79]. Figure 1.8 visualizes a sample of friends in New York City. The red links between users represent friendships, and the colored pins show their current geographical locations on a map. The highlighted person is complaining about her health, and

---

[7]http://livehoods.org/
[8]http://tweettracker.fulton.asu.edu/

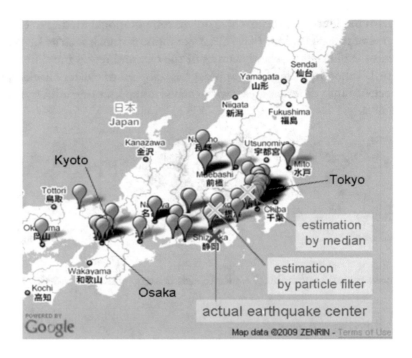

Figure 1.6: An estimation of affected areas during Japan earthquake in August 2009 (based on ([80])).

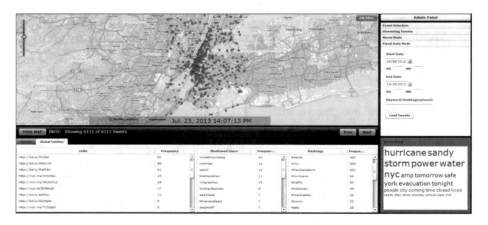

Figure 1.7: TweetTracker for disaster relief (based on ([50])). Tweets are organized into events and visualized on the crisis map to facilitate disaster relief with search and analysis tools.

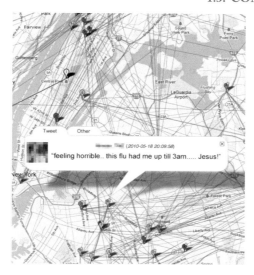

Figure 1.8: An example of disease spread (based on ([79]). The links between users represent friendships, and the pins show their geographical locations. The highlighted person complains about her health and hints the symptom. Disease spread detection systems investigate the impact of such a person on the health of her friends and people around her.

hinting about the specifics of her ailment. The system investigates the impact of such a person on the health of his friends, and of people around him.

Mobile Security and Privacy

Location sharing is a fundamental function of location-based social networking services. Users share their locations by checking in on location-based social networking sites to let their friends know where and when they are. The location awareness can then form LBSNs and enhance the user's social connections. For example, a user may want to hang out with his friend after knowing that he is nearby through his check-in status. While location sharing significantly enhances user experience in social networks, it also leads to privacy and security concerns. In recent years, location privacy on LBSNs has attracted more and more attention from both academia and industry. It has been discovered that privacy is a critical concern for user when considering adopting location-sharing services [18, 37, 52, 96]. When using location-sharing services, some users would like to share their location with friends for social purposes, while other users may believe that sharing personal location discloses one's personal preferences, which may cause potential physical security risks. Therefore, it is inevitable to consider privacy management when designing location-sharing applications.

To improve privacy management on location-based services, one of the major challenges is to understand why people are using location-sharing services and under what circumstances they do not want to share locations. It has been analyzed on Dodgeball that location-based social services influence the way people experience urban public places and their social relations [45]. It has also been explored on how and why people use Foursquare w.r.t. five factors, i.e., badges and fun, social connection, place discovery, keeping track of places, and competition with themselves [60]. Generally, the majority of users had few privacy concerns, and users chose not to check in at specific locations mainly because the places were embarrassing, non-interesting, or sensitive. Mobile applications have also been developed to help manage privacy on LBSNs. "Locaccino"[9] is a location-sharing application focusing on privacy control based on the Facebook social network [77, 95]. A Locaccino user can request the location of his Facebook friends. It allows a user to set detailed location-sharing privacy preferences such as when and where his location can be visible to a set of pre-specified users.

1.3.2   HUMAN MOBILITY: REPETITIVE VS. COLD-START

The above section highlights several research and application opportunities for studying human mobility from different perspectives with large-scale LBSN data. Since location is the most critical element in reflecting a user's mobility patterns. To study human mobility, it is inevitable to start from the location aspect w.r.t. check-in actions.

One of the most significant properties of check-in behavior is the user-driven property [73]. When using location-based social networking services, a user can choose where and when to check-in. It has been discovered that a user's check-ins follow a power-law distribution on LBSNs, i.e., a user goes to a few places many times and to many places a few times [28], indicating that users would like to: (1) return to visited locations, resulting in the repetitive check-in behavior; or (2) explore new locations, resulting in the cold-start check-in behavior.

Predicting whether a user would like to return to a previously visited location is generally corresponding to a computational task on LBSNs, "location prediction." Various mobility patterns discovered from LBSN data, such as spatial trajectories [71, 87], periodic patterns [94], and spatio-temporal patterns [29, 82], are leveraged for this task. On the other hand, when a user explores a city and wants to try new points of interest, "location recommendation" is the task that can facilitate users on such purpose. In this book, we focus on the above two basic types of human mobility patterns: returning to visited locations and exploring new locations to visit.

The remainder of this book consists of three chapters. Chapter 2 discusses the basic data structures and properties on LBSN data, together with the fundamental discoveries of human mobility patterns exhibited on LBSNs. Chapter 3 introduces methodologies of location prediction for mining human mobility w.r.t. determining which visited location a user would like to return. Chapter 4 expands further on location recommendation approaches and discuss how to facilitate

---

[9]http://locaccino.org/

a user's exploration on new locations through the analysis of his LBSN contexts. In Chapter 5, we discuss several topics that are beyond the Human Mobility, such as location privacy issues.

CHAPTER 2

Analyzing LBSN Data

Location-based social network data contains various types of information, providing us with unprecedented opportunities to study human mobility. In this chapter, we introduce the properties of data collected from typical LBSN websites, and human mobility patterns.

2.1   A CHECK-IN EXAMPLE

Location-based social network (LBSN) data is generated through users' check-ins. When a user performs a check-in action through LBSN services, it generates a check-in record. Figure 2.1 shows a typical check-in record, where four "**W**" elements are present.

Figure 2.1: A check-in example with four "**W**" elements.

- **Who**
  A check-in contains a "who" element, e.g., "Felix" in this example. *Social information* is also present, i.e., "Jiliang" is a friend of "Felix."

- **When**
  The time of the check-in tells us "when" it happens, consisting of the *temporal information*, e.g., "August 1, 2013."

- **Where**

  The location of a check-in action shows "where" it happens, indicating the *geographical information*, e.g., "Shanghai Flavor Shop" in "Sunnyvale, CA."

- **What**

  The content such as comments, tips, and location descriptive tags presents "what" the users do at this location, corresponding to the *content information*, e.g., "Best pan-fried pork bun and Shanghai wonton on the west coast." "Liking" indicates the user's sentiment regarding the location.

The above example presents the key elements of a check-in action. More information could be obtained through the APIs of the LBSN service provider, e.g., accurate geographical position, location descriptive tags, user tips, list of social friends, etc.

Below we introduce in detail the structure of typical LBSN data.

2.2   STRUCTURE OF LBSN DATA

Typically, LBSN data can be organized in four types regarding users' check-ins, user profiles, social networks, and location descriptions, as shown in Figure 2.2.

- **Check-ins**

  The check-in action table contains records of check-ins. Each row represents a check-in action containing user, check-in location, check-in time, and user-generated content, such as comments, tips, or likes if available. The example in Figure 2.2 shows two check-in actions. Depending on different location-based social networking services, information may vary.

- **User Profiles**

  A user profile contains a user's personal information, including his home location region, personal interests described by tags, etc., as shown in Figure 2.2. This part of information is generally private and can be determined by the user through LBSN websites regarding how much to release to the public.

- **Social Networks**

  Users' social information is stored in a social network table, with each row representing a connection between two users. The example in Figure 2.2 lists the social connections of user #505 and #399. Generally, social connections on LBSNs are undirected.

- **Location Descriptions**

  The check-in locations are stored in the location description table. Each row contains information such as name, latitude/longitude, address, category, rating, and tags.

This data structure is commonly seen at many LBSN websites, including Foursquare, Yelp, Gowalla, and BrightKite. Table 2.1 lists some typical LBSN datasets that are available for re-

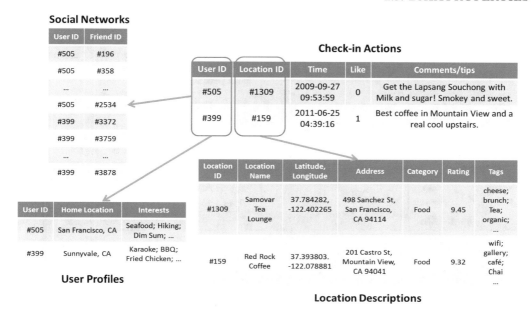

Figure 2.2: The structure of LBSN data contains users' check-ins, user profiles, social networks, and location descriptions. The users and locations are represented as anonymous IDs.

search. Some related projects are shown together with these datasets as they used the corresponding datasets. These datasets are all anonymized for privacy concerns.

We introduce next the LBSN data properties.

2.3 DATA PROPERTIES

Compared to traditional cellphone data from GPS trackers or telecommunication companies, LBSN data has three distinct properties.

2.3.1 SOCIO-SPATIAL PROPERTIES

Geographical properties and social connections are two special factors on LBSNs. The geographical properties reflect human behavior in the real world, distinguishing location-based social networks from content-based social networks [81]. The explicit social network information, which is generated by users through "add friend," distinguishes location-based social networks from cellphone data. On cellphone-based GPS data, social information is commonly collected via user study [21, 54] or inferred from communication networks through calling/messaging records or bluetooth connections [97].

Table 2.1: Location-based social networks datasets

Dataset	Duration	Download
Geolife [116]	Apr. 2007 - Aug. 2012	`http://research.microsoft.com/en-us/projects/geolife/`
Gowalla [17]	Feb. 2009 - Oct. 2010	`http://snap.stanford.edu/data/loc-gowalla.html`
Brightkite [98]	Apr. 2008 - Oct. 2010	`http://snap.stanford.edu/data/loc-brightkite.html/`
Twitter [16]	Sep. 2010 - Jan. 2011	`http://infolab.tamu.edu/data/`
Yelp [61]	Oct. 2004 - Jan. 2015	`http://www.yelp.com/dataset_challenge/`
TweetTracker [50]	Since Oct. 2010	`http://tweettracker.fulton.asu.edu//`

Furthermore, social networks and geographical properties, are two correlated elements that affect each other [100, 111]. For example, a user is more likely to befriend other users who are geographically close, e.g., co-workers, colleagues. Likewise, a user may check-in at a location due to the influence from his friends, such as following friends' suggestions to visit a restaurant, going out with friends for shopping. These correlations result in a new property, commonly referred to as socio-spatial properties.

Figures 2.3 and 2.4 present the probability of a social tie between two users w.r.t. their geographical distance on three typical LBSN datasets. It shows that users with short geographical distance are more likely to become friends than users with long geographical distance, and friends usually have short geographical distance than non-friends [3, 83].

2.3.2   LARGE-SCALE AND SPARSE DATA

In traditional cellphone-based GPS data, users' geographical movements are automatically tracked and recorded by the service provider according to a pre-defined time interval (usually within 10 min). The increasing use of mobile devices has led to the availability of big mobile data. Thus, one can easily obtain a long trajectory with massive GPS points from a single user. However, studies on such data are still limited to the small number of participant users [21, 118].

In contrast, location-based social networks adopt a user-driven check-in strategy [73], i.e., the user decides whether or not to check-in at a specific place based on his own choices. For example, a user may check-in at a museum after his last check-in at a restaurant two days earlier. Hence, check-in data on LBSNs can also be **sparse**, which greatly increases the difficulty of mining human mobility. LBSN data is also **big**. Different from the limited participated users in traditional GPS data, location-based social networking services take advantage of the social media platform, generating a large number of geographical check-ins from millions of users [11, 83].

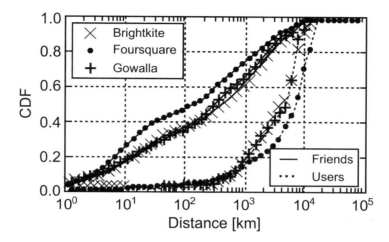

Figure 2.3: Empirical Cumulative Distribution (CDF) of the geographic distance between all users and between connected friends for three LBSN datasets of Brightkite, Foursquare, and Gowalla, respectively (based on ([83])).

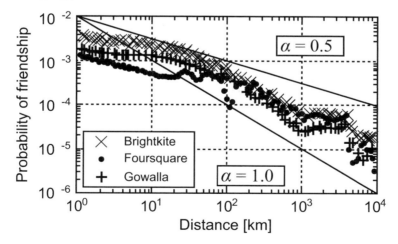

Figure 2.4: Probability of friendship between two users as a function of their geographic distance for the three datasets (based on ([83])).

For example, Foursquare reached 55 million users and over 6 billion check-ins by November, 2014.[1] Yelp announced a monthly average of 132 million unique visitors in Q1 2014, and more than 57 million reviews by the end of Q1 2014.[2] Figure 2.5 presents the user distributions on LBSNs w.r.t. the world and the U.S., respectively.

2.3.3 SEMANTIC INDICATION

Location-based social networking services provide semantic indications on check-in locations. Traditional cellphone data stores GPS trajectories in terms of longitude and latitude, and there is generally no mapping between a GPS position to a specific location (i.e., a POI). For example, it is not easy to figure out whether a GPS point corresponds to a restaurant, theater, or just point on highway. Although there are location databases available aimed at generating such correlations, there are still unsolved problems. First, some locations may share the same (or close) GPS positions, such as two adjacent POIs on the same street, or two businesses share the same location and one is upstairs of the other. Second, even if the GPS position can be accurately mapped to a POI, it is still unsure whether the user has really visited that location. The observation of a user staying at a GPS position close to a book store for 10 min does not necessarily indicate that the user has visited that store.

Data generated on LBSNs is user-driven. The above problems are naturally addressed. A location is associated with a user's check-in only if the user specifies it through LBSN services, thus solving the uncertainty of the check-in locations as with traditional cellphone data. When a user plans to check-in, LBSN services provide him with nearby POIs according to the user's current position. The user selects the correct one and performs a check-in action by clicking the "check-in" button for the location. Each location in LBSN is associated with textual descriptions such as tags, categories, names, as shown in Figure 2.2. Furthermore, the user can provide additional content of the location in terms of tips and comments. Figure 2.6 plots the check-in distribution over different geographical feature types such as restaurants and bars. Figure 2.7 presents the semantic information from worldwide check-ins on a word cloud generated using tags. Such semantic information helps understand LBSN users' mobile preferences [55, 56].

2.4   MOBILITY PATTERNS

Mobility patterns can be discovered from LBSN data. In this section, we discuss some representative mobility patterns regarding their properties, which will be further used for mining human mobility in Chapters 3 and 4.

---

[1]https://foursquare.com/about
[2]http://www.yelp-press.com/phoenix.zhtml?c=250809&p=irol-press

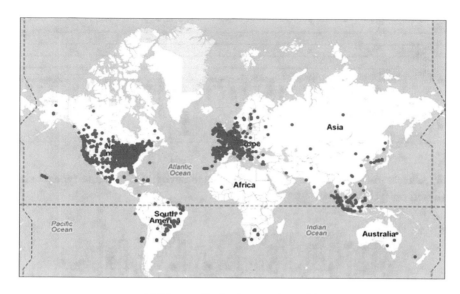

(a) The user distribution over the world.

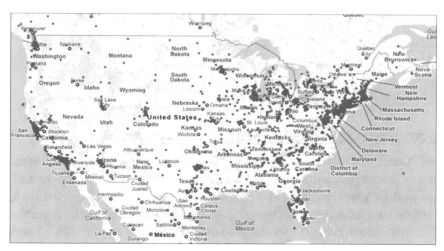

(b) The user distribution over the USA.

Figure 2.5: The user distribution on Foursquare (based on ([31])).

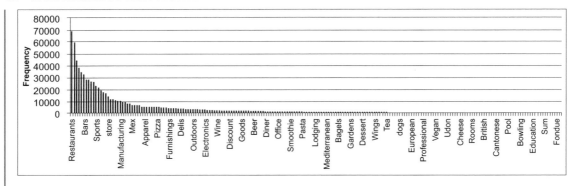

Figure 2.6: Check-in frequency distribution for selected geographic feature types such as restaurants and bars (based on ([102])).

Figure 2.7: Semantic information from worldwide check-ins (based on ([16])).

2.4.1 INVERSE DISTANCE RULE

The idea of "Death of Distance" proposed in 2011 claims that geographical distance plays a less important role due to the communication revolution and the rapid development of the Internet, making our world a "global village" [9]. Studies on spatial structures of networks demonstrate that there is a strong correlation between geographical attributes and network properties, and suggest the significance of considering the spatial properties in network analysis [33]. Various studies argue that the IT revolution does not turn us into a borderless society, as physical proximity still plays an important role in the Internet era [34, 70]. One of the first attempts to investigate how social connection is affected by geographical distance in online social networks was conducted on LiveJournal [59]. Analysis on users' social networks and their hometown information shows that only one-third of friendships are independent of geography. In addition, studies on the first commercial LBSN service in the U.S., Dodgeball, indicate that locations do change people's online experience [45].

The increasing use of location-based social networking services generates large-scale data with geographical distance between users and their social networks. The study on geo-social metrics [81] reports that: (1) users who live close to each other have a higher probability of creating friendship links than those who are distanced; and (2) users in the same social cluster show short geographical distances. Furthermore, by comparing location-based social networks (Brightkite and Foursquare) with content-based social networks (LiveJournal and Twitter), it is shown that people within a social cluster on LBSNs tend to have smaller geographical distance than online social networks that focus on content producing and sharing.

Researchers also investigated how geographical distance influences social networks, and how social networks influence human movement on LBSNs. One study on three location-based social networking sites (Brightkite, Foursquare, and Gowalla) [83] discovers strong heterogeneity across users at different geographic scales of interactions across social ties. The probability of a social tie between two users is roughly a function of the geographical distance between them. The study on LBSN data and cell phone data [17] reports that long-distance travel is more influenced by social friendship while short-range human movement is not influenced by social networks. The investigation [49] on Twitter social networks concludes that offline geography still matters in online social networks, while one third of the users would like to have their social links in other countries, which is consistent with the previous findings in [59, 81]. Figures 2.4 and 2.8 present such relationships.

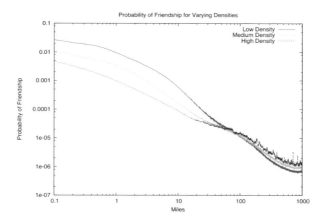

Figure 2.8: Looking at the people living in low-, medium- and high-density regions separately, we see that if one lives in a high-density region (a city), he is less likely to know nearby individuals, since there are so many of them (based on ([3])).

2.4.2   LÉVY FLIGHT OF CHECK-INS

"Lévy Flight" property indicates that people tend to visit nearby places and occasionally distant places. The study on millions of check-ins on Facebook [76] discovers that "Lévy Flight" does exist on LBSNs. Similarly on Foursquare, around 20% of consecutive check-ins in Foursquare happen within 1 km of one another, 60% between 1 and 10 km, and 20% over 10 km [73]. Figure 2.9 plots the Lévy Flight of check-in data for the distance between two continuous check-ins, while Figure 2.10 presents a similar plot for the distance between a user's check-in location to his home. In reality, such a property can be used to predict a user's location or recommend locations to him. For example, by investigating the "Lévy Flight" patterns of a user's check-in history, one can generate a short list of location candidates for a recommendation based on the user's current position.

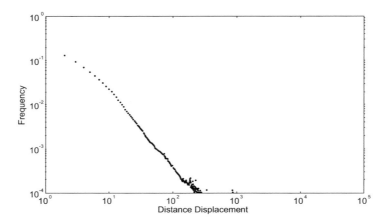

Figure 2.9: Lévy Flight of check-in data for consecutive check-ins (based on ([16])).

2.4.3   POWER-LAW DISTRIBUTION AND SHORT-TERM EFFECT

The "Lévy Flight" property describes mobility patterns between two check-ins. One can organize a set of check-ins according to their time stamps, and generate a historical check-in sequence. We discuss two properties observed from such check-in sequences on LBSNs. First, a user's check-in sequence follows power-law distribution, i.e., a user goes to a few places many times and to many places a few times. Figure 2.11(a) shows the distribution of check-in frequency on a LBSN data. Note that both x-axis and y-axis are in the log scale. The figure suggests that the check-in sequence follows a power-law distribution

$$N(x) \approx a \times x^{b}, \tag{2.1}$$

where x is the check-in frequency, $N(x)$ is the number of locations with check-in frequency as x, and b is the exponent which is approximately 1.42. The check-in distribution of an individual also shows power-law property, as shown in Figure 2.11(b).

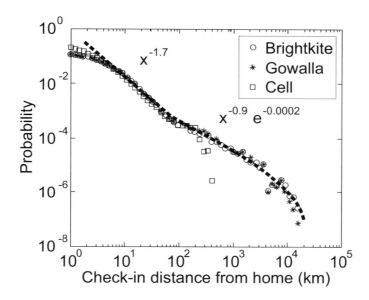

Figure 2.10: Lévy Flight of check-in data between a user's check-in location to his home (based on ([17])).

Second, the check-in actions in a sequence have a short-term effect. For example, considering that a user arrives at the airport and then takes a shuttle to the hotel, has his dinner and sips a cup of coffee. The effects of the previous check-ins at airport, shuttle stop, hotel, and restaurant have different correlation strength to the latest check-in of drinking coffee. Such effect decreases as the time goes on.

The similar structures shared by location trajectories and word sequences are shown in Table 2.2. For example, a document in language processing can correspond to an individual check-in sequence in LBSNs, while a word in the sentence corresponds to a check-in location. Thus, the power-law distribution and short-term effect are shared by LBSNs and natural language processing, where the word distribution is closely approximated by the power-law [121]; and the current word is more relevant to its adjacent words than distant ones. The language model that works with word sequences is potentially applicable to location trajectories on LBSNs due to these common features. For example, the unigram language model that ignores the relationship between the current word to its nearest neighbors can be applied to LBSNs while considering the current check-in and ignoring its nearby check-ins, and so can the n-gram language model. In

Chapter 3, we introduce how to predict a visited location by exploring these two properties with methods derived from language models.

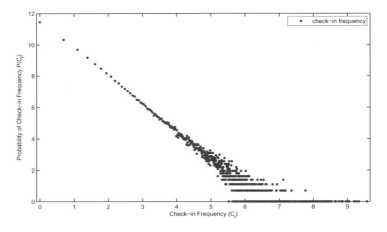

(a) Power-law distribution of check-ins in whole dataset

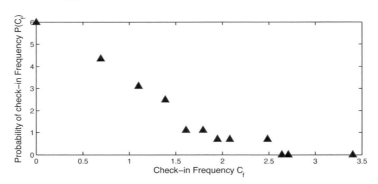

(b) Power-law distribution of individual check-ins

Figure 2.11: The power-law distribution of check-ins (based on ([28])).

2.4.4   TEMPORAL PERIODIC PATTERNS

Human mobility exhibits significant temporal periodic patterns on LBSNs, which are correlated to the location property, such as daily patterns (hours of the day), weekly patterns (days of the week), and weekday/weekend patterns [16, 53, 67, 73, 102]. For example, a user may regularly arrive to the office around 9:00 am, go to a restaurant for lunch at 12:00 pm, and watch movies at night around 10:00 pm. Therefore, investigating the features embedded in daily patterns allows us to better understand human mobility, providing a potential opportunity to design more advanced location-based services on LBSNs.

Table 2.2: Common properties of language processing and LBSNs mining ([28])

Language processing		LBSNs mining	
Corpus		Check-in collection	
Document		Individual check-ins	
	Paragraph		Monthly check-in sequence
Document Structure	Sentence	Check-in Structure	Weekly check-in sequence
	Phrase		Daily check-in sequence
	Word		Check-in location

Figures 2.12 and 2.13 present the users' daily check-in behavior in the world and in three cities. Figure 2.14 presents the weekly check-in behavior globally. These plots indicate that there are certain time periods when users would like to visit locations, considered as "peak," and certain time periods when users rarely visit locations, considered as "bottom." As these patterns are about human mobility in temporal aspects, we will discuss in detail in later sections.

2.4.5 MULTI-CENTER CHECK-IN DISTRIBUTION

Multi-center check-in distribution has been observed from large-scale LBSN data. The "center" could be a geographical area, or a specific time period. From a spatial aspect, a user's movement generally centers on certain location areas, and rarely checks-in at locations far away from the center. Furthermore, the check-in probability follows the inverse distance rule (i.e., Lévy Flight). Figures 2.15 and 2.16 plot a user's check-in history centered at two points "home" and "work."

From a temporal aspect, a user tends to visit a location at certain time periods, such as visit a restaurant between 11:00 am to 1:00 pm, and 5:00 pm to 7:00 pm, and rarely visit the location at other time periods. Figures 2.17 and 2.18 plot a user's daily and weekly check-in distribution at a location l from a LBSN dataset. Each point represents the total number of check-ins at a specific hour of the day (day of the week) at location l by that user, respectively. It can be observed that the check-in probability is centering on certain time periods and decreasing during other time periods.

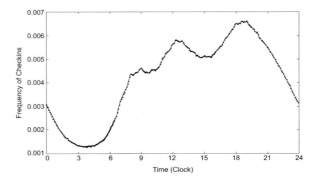

Figure 2.12: Temporal distribution of check-in data in the world (daily patterns) (based on ([16])).

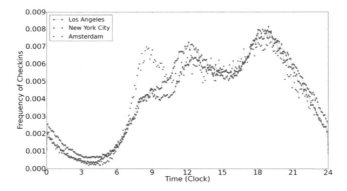

Figure 2.13: Temporal distribution of check-in data in three cities (daily patterns) (based on ([16])).

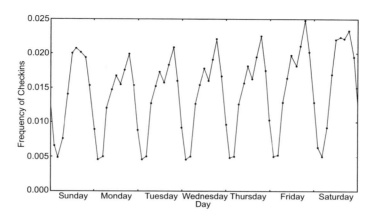

Figure 2.14: Temporal distribution of check-in data in the world (weekly patterns) (based on ([16])).

Figure 2.15: Check-ins of a user in San Francisco: geographic distribution of check-ins when at home or work. (based on ([17])).

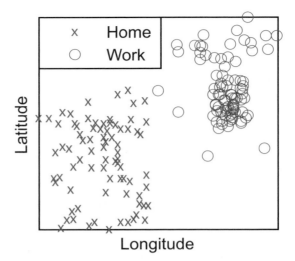

Figure 2.16: Check-in locations generated by home/work states (based on ([17])).

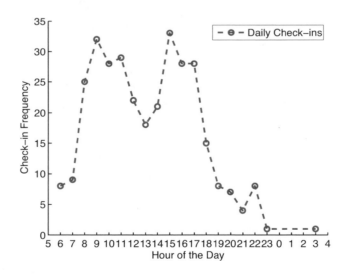

Figure 2.17: Daily check-in distribution of a user at location l (based on ([26])).

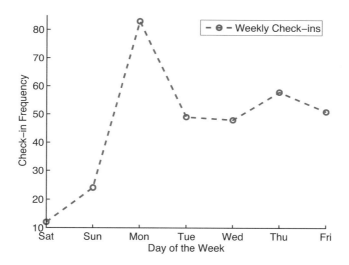

Figure 2.18: Weekly check-in distribution of a user at location l (based on ([26])).

CHAPTER 3

Returning to Visited Locations

All of us return to previously visited locations, e.g., offices, homes, restaurants, shopping stores, or nearby movie theaters. Capturing such behavior corresponds to a computational task called *location prediction*. Location prediction is a traditional task in mobile computing of discovering users' repetitive mobility patterns. Observing a user's visiting history can help discover the user's interests in different visited locations. Predicting which location the user would most likely visit again could promote mobile marketing and facilitate users' daily life, such as distributing relevant restaurant coupons to a user based on the prediction.

Location prediction on LBSNs refers to two different prediction targets: (1) predicting a user's next visiting location and (2) predicting a user's home location. Due to the user-driven check-in property, a user tends to check-in at certain "exciting" locations, but rarely check-in at his home location, not only because the latter is less "exciting," but also because of certain privacy concerns. Thus, the prediction of a user's next visiting location focuses on those "exciting" locations visited by the user, which relies on discovering of the mobility patterns through a user's visiting history from spatial, temporal, and social aspects. On the other hand, the prediction of a user's home location usually relies on content information and network information, as few visiting records of home location can be observed from a user's visiting history. In this chapter, we introduce definitions and present representative location prediction approaches in each category.

3.1 NEXT LOCATION PREDICTION

Next location prediction is to predict a user's next visiting location based on the user's check-in history on LBSNs. Next location prediction algorithms are discussed in terms of four types: *sequential patterns*, *temporal dynamics*, *social correlations*, and *hybrid patterns*.

3.1.1 SEQUENTIAL PATTERNS

Sequential patterns suggest that a user follows some sequential behaviors when visiting locations. Consider a user's check-in history as a sequence of locations. Prediction is made based on the count of each location appeared in the sequence, and its relationship to nearby locations. Prediction algorithms with sequential patterns usually consider two factors: *check-in history* and *check-in context*. The *check-in history* refers to a user's historical check-ins during a certain period, e.g., last year. It is used to discover sequential patterns of the user. The *check-in context* is a check-in sequence extracted from the user's latest check-in locations. According to how many latest check-ins are considered as context, methods in this area can be classified as *Most Frequented*

Check-in, Order-k Markov, Fallback Markov, and *Combined Models*. Note that the *check-in history* may cover the *check-in context* if the former contains a user's latest check-ins. For ease of illustration, we consider them as two independent factors in this chapter.

Most Frequented Check-in Model

The Most Frequented Check-in (MFC) model is mostly applied when there is no observed *check-in context* but only the observed *check-in history*. It assigns the probability of next check-in c_n at location l as the probability of l appearing in the check-in history, the underlying hypothesis being that a user tends to visit his favorite locations.

$$P_{MFC}(c_n = l|\mathcal{H}_u) = \frac{|\{c_i \mid c_i \in \mathcal{H}_u, c_i = l\}|}{|\mathcal{H}_u|}, \tag{3.1}$$

where $\mathcal{H}_u = \{c_1, c_2, ..., c_N\}$ is the set of historical check-in locations of user u.

Example 3.1   *Given a user's check-in history for one week, as shown in Table 3.1, assume the user is going to check-in at a location now, and predict this location using the Most Frequented Check-in model.*

According to the time stamps, the check-in locations can be chronologically organized as the following location sequence:

$$\mathcal{H}_u = \{l_1, l_2, l_3, l_2, l_3, l_4, l_5, l_1, l_3, l_2, l_6, l_2, l_3, l_2, l_3, l_2, l_6, l_2, l_5, l_3\}.$$

With MFC, we find the most frequently visited location in $\mathcal{H}_u$ as shown in Table 3.2, which presents the statistical information of each location in $\mathcal{H}_u$.

Equation (3.1) shows how to estimate the check-in probability. For example, the probability of the next check-in happening at l_1 is $2/20 = 0.10$, where 20 is the total number of check-ins in the user's check-in history. From Table 3.2, we see that l_2 is visited most. Thus, MFC predicts the next visiting location as l_2.

Order-*k* Markov Model

The Order-*k* Markov (OM-k) model [86] is an approach to mine associated patterns from sequential data. It is mostly applied when both *check-in context* and *check-in history* are observable. The OM-k considers the latest k visited locations as check-in context $\mathcal{C}_u$, and searches for frequent patterns in the check-in history $\mathcal{H}_u$ to predict the next location. The probability of a location l being visited is defined by the Order-*k* Markov model as:

$$\begin{aligned} &P_{OM-k}(c_n = l|\mathcal{H}_u, \mathcal{C}_u) \\ &= \frac{|\{c_i \mid c_i \in \mathcal{H}_u, c_i = l, c_{i-j} = c_{n-j}\}|}{|\{c_i \mid c_i \in \mathcal{H}_u, c_{i-j} = c_{n-j}\}|}, 0 < j \le k, j \in \mathbb{Z}, \end{aligned} \tag{3.2}$$

where $\mathcal{C}_u = \{c_{n-k}, ..., c_{n-2}, c_{n-1}\}$ is the check-in context consisting of the latest k check-ins of user u.

Table 3.1: An example of check-in history from an individual user

ID	Time	Location
1	10:25 am, Sun, Mar 16, 2014	l_1
2	11:12 am, Sun, Mar 16, 2014	l_2
3	01:30 pm, Sun, Mar 16, 2014	l_3
4	20:16 pm, Sun, Mar 16, 2014	l_2
5	22:50 pm, Sun, Mar 16, 2014	l_3
6	10:30 am, Mon, Mar 17, 2014	l_4
7	11:26 am, Tue, Mar 18, 2014	l_5
8	11:56 am, Tue, Mar 18, 2014	l_1
9	01:30 pm, Tue, Mar 18, 2014	l_3
10	11:02 am, Wed, Mar 19, 2014	l_2
11	11:55 am, Wed, Mar 19, 2014	l_6
12	11:08 am, Thu, Mar 20, 2014	l_2
13	01:30 pm, Thu, Mar 20, 2014	l_3
14	11:21 am, Fri, Mar 21, 2014	l_2
15	11:20 pm, Fri, Mar 21, 2014	l_3
16	00:30 am, Sat, Mar 22, 2014	l_2
17	11:52 am, Sat, Mar 22, 2014	l_6
18	01:30 pm, Sat, Mar 22, 2014	l_2
19	07:22 pm, Sat, Mar 22, 2014	l_5
20	11:30 pm, Sat, Mar 22, 2014	l_3

Table 3.2: Predicting the next location with the Most Frequent Check-in model

Location	l_1	l_2	l_3	l_4	l_5	l_6
No. of Visits	2	7	6	1	2	2
Check-in Probability	0.10	0.35	0.30	0.05	0.10	0.10

Example 3.2 *Given a user's check-in history in Table 3.2, assume that his latest check-ins are $l_1 \rightarrow l_3 \rightarrow l_2$ (not listed in Table 3.2), and predict his next visiting location with Order-k Markov model.*

An Order-k Markov model considers the latest k check-ins as check-in context. First, we need to determine k. If $k = 1$, we use the latest check-in as check-in context, i.e., l_2, and search for sequential patterns starting with this context, i.e., $l_2 \rightarrow x$, where x is a check-in location happened after the context. If $k = 2$, we consider the last two check-ins as context, i.e., $l_3 \rightarrow l_2$, and search for sequential *patterns starting with these two check-in locations, i.e., $l_3 \rightarrow l_2 \rightarrow x$. Table 3.2 presents the statistical information of each order in $\mathcal{H}_u$.*

Table 3.3: Predicting the next location with the Order-k Markov model

Location	k=1			k=2	
Sequences	$l_2 \rightarrow l_3$	$l_2 \rightarrow l_6$	$l_2 \rightarrow l_5$	$l_3 \rightarrow l_2 \rightarrow l_3$	$l_3 \rightarrow l_2 \rightarrow l_6$
No. of Visits	4	2	1	2	1
Check-in Probability	0.57	0.29	0.14	0.67	0.33

The check-in probability is computed as in Eq. (3.2). For example, the probability of visiting l_3 after visiting l_2 is $4/7 = 0.57$, where 7 is the total number of length 2 patterns starting with l_2. When $k = 1$, according to Table 3.3, we see that after visiting l_2, there is 57% probability for the user to visit l_3, 29% probability to visit l_6, and 14% probability to visit l_5. Thus, the predicted next visiting location is l_3 with the Order-1 Markov model. Similarly, the most likely visited location after consequently visiting l_3 and l_2 is l_3 (67%). Thus, the predicted next visiting location is also l_3 with the Order-2 Markov model.

Fallback Markov Model

The MFC Model is actually an Order-0 Markov Model without considering any latest check-in locations as context. One problem of the Order-k ($k \geq 1$) Markov model is that it relies on the observable context. Thus, if the context is not in the check-in history, the model would fail immediately. For example, if the check-in context is $C_u = \{l_1, l_7, l_2\}$ and the Order-2 Markov model is chosen, according to the check-in history $\mathcal{H}_u$ in Example. 3.1, it is the first time this user checks-in at l_7, while no sequential patterns for the order-2 pattern $l_7 \rightarrow l_2 \rightarrow x$ can be extracted from $\mathcal{H}_u$. Therefore, the Order-2 Markov Model is not applicable to predict the next location. Actually, any pattern with k larger than 2 in this example would fail, and only the Order-1 Markov model or MFC model could work. The Fallback Markov (FM) model is then proposed.

In the Fallback Markov model, a bigger k is first selected, say, $k = 3$. Considering the check-in context $C_u = \{l_1, l_7, l_2\}$, the sequential patterns regarding $k = 3$ are searched from the check-in history, which corresponds to $l_1 \rightarrow l_7 \rightarrow l_2 \rightarrow x$ pattern in $\mathcal{H}$, where x represents any location satisfying that pattern. Since there is no such pattern in $\mathcal{H}$, the model falls back to the Order-2 Markov model ($k = 2$), and search for order-2 patterns $l_7 \rightarrow l_2 \rightarrow x$. This process goes on until an order-k pattern is found. In an extreme case when $k = 0$, the MFC model is adopted and the most frequently visited location in $\mathcal{H}$ is considered for prediction. The reason to start from a big k is because that bigger k indicates longer patterns, and the probability of a long pattern being repeated is relatively smaller than a short pattern; thus, a long pattern generally provides strong indications on mobility patterns, while the fallback model tends to consider strong indications first.

Combined Markov Model

The above Markov models (i.e., OM and FM) need to specify a parameter k, which considers only one context at a time. However, the choice of k commonly requires human knowledge, and different order-k patterns may contain complementary information for prediction. Thus, combined models are developed to address this issue. In a combined model, there is no need to specify a k for the Markov model. It considers order-k sequential patterns with different k, and combines them together with smoothing which assigns weights on each pattern. The following example illustrates the basic idea of Combined Markov (CM) model.

Example 3.3   *Given a user's check-in history in Example 3.1, and his latest check-in context* $C_u = \{l_1 \rightarrow l_3 \rightarrow l_2\}$, *predict his next visiting location with the Combined Markov model.*

Table 3.4 lists the probability of next visiting location based on the current context with different k. *For example, the probability of visiting* l_3 *with latest check-in context as* $\{l_3 \rightarrow l_2\}$ *is 0.67, while the probability of visiting* l_4 *with latest check-in context as* $\{l_1 \rightarrow l_3 \rightarrow l_2\}$ *is 0.*

Table 3.4: Probability of the next location with different contexts

		Probability of Next Visiting Location					
	Context	l_1	l_2	l_3	l_4	l_5	l_6
k=0	N/A	0.10	0.35	0.30	0.05	0.10	0.10
k=1	l_2	0	0	0.57	0	0.14	0.29
k=2	$l_3 \rightarrow l_2$	0	0	0.67	0	0	0.33
k=3	$l_1 \rightarrow l_3 \rightarrow l_2$	0	0	0	0	0	1

Assume the weight of each order-k pattern is given in Table 3.5, and the final probability of visiting a location is a weighted combination on these order-k patterns:

$$P(l) = \sum_k P_k(l) * w_k, \tag{3.3}$$

where $P_k(l)$ *is the probability of location l with order-k context, and* w_k *is the weight of corresponding order-k context. For example, the combinational probability of visiting* l_3 *is*

$$P(l_3) = 0.30 * 0.40 + 0.57 * 0.30 + 0.67 * 0.10 + 0 * 0.20 = 0.3580 \tag{3.4}$$

and the combinational probability of visiting l_6 *is*

$$P(l_6) = 0.10 * 0.40 + 0.29 * 0.30 + 0.33 * 0.10 + 1 * 0.20 = 0.3600. \tag{3.5}$$

A similar procedure can be applied to calculate the probability of other locations. The final prediction of the next visiting location in this example is l_6.

Table 3.5: Predicting the next location with the Combinational Markov model

Location	$k = 0$	$k = 1$	$k = 2$	$k = 3$
Context	N/A	l_2	$l_3 \rightarrow l_2$	$l_1 \rightarrow l_3 \rightarrow l_2$
Weight	0.40	0.30	0.10	0.20

In this example, weights are pre-defined for ease of presentation. In practice, the weight of each order-k context is commonly learned from the *check-in history* by mathematical models. These models originate from n-gram language modeling, such as the Interpolated Kneser-Ney or hierarchical Pitman-Yor language model. Since it has been discovered that word sequence and location sequence share a set of common properties, i.e., power-law distribution and short-term effect [28] discussed in Section 2.4.3, these language models work well when applied to location sequence for location prediction. The learning process with these methods is beyond the scope of this book. For detailed information readers can refer to [93].

3.1.2   TEMPORAL DYNAMICS

The sequential patterns discussed above are generally for next location prediction if we do have information of a user's next visiting time. If such information is available, temporal patterns can be used. Temporal periodic patterns from a user's repetitive check-in behavior can be useful in next location prediction.

Most Frequented Time Model

Human geographical movement exhibits strong temporal patterns and is highly relevant to the location property [16, 67, 102]. As discussed in Section 2.4.4, people can have repetitive check-ins in daily activities such as regularly going to a restaurant for lunch around 12:00 pm, watching movies on Friday night, and shopping during weekends. Various temporal periodical patterns, such as daily pattern (hour of the day) and weekly pattern (day of the week), can be discovered through this kind of check-in behavior. We first introduce an intuitive model, the Most Frequented Time (MFT) model.

We start from hours of the day patterns to illustrate the MFT model. Let h_i denote the hourly information of the check-in c_i, where $h_i \in \{0, 1, ..., 23\}$ indicating one of the 24 h. For example, the third check-in with time "1:30 pm, Sunday, Mar 16, 2014" in Table 3.1 corresponds to $h = 13$. $\mathcal{H}_u$ is the observed historical check-ins of user u. The MFT model assigns the probability of the check-in c_n at location l at time h_n as the probability of location l occurring at hour h_n in the previous check-in history,

$$P_{MFT}(c_n = l | h_n, \mathcal{H}_u) = \frac{|\{c_i \mid c_i \in \mathcal{H}_u, c_i = l, h_i = h_n\}|}{|\{c_i \mid c_i \in \mathcal{H}_u, h_i = h_n\}|}. \quad (3.6)$$

Example 3.4   *Given a user's check-in history in Table 3.1, assume that the user is going to check-in at a location tomorrow around 11:30 am; predict that location with the Most Frequented Time model.*

The time information "11:30 am" corresponds to $h = 11$. In Table 3.1, there are 8 check-ins that happen at $h = 11$, whose check-in IDs are 2, 4, 8, 11, 12, 14, 17, and 19. The corresponding check-in locations are $\{l_2, l_2, l_1, l_6, l_2, l_2, l_6, l_5\}$. Table 3.2 presents the probability of each location being visited at the given time.

Table 3.6: Predicting the next location with the MFT model (daily patterns)

Location	l_1	l_2	l_3	l_4	l_5	l_6
No. of Visits	1	4	0	0	1	2
Check-in Probability	0.125	0.50	0	0	0.125	0.25

The check-in count is computed as in Eq. (3.6). For example, the probability of visiting l_6 is $2/8 = 0.25$, where 8 is the total number of check-ins made by the user at $h = 11$ in Table 3.1. According to Table 3.6, there is 50% probability for the user to visit l_2 at 11:30 am, which is higher than the probabilities of all the other location candidates. Thus, the predicted location is l_2 based on the MFT model with the consideration of hour of the day patterns.

Similarly, for weekly patterns, we introduce $d \in \{1, 2, 3, 4, 5, 6, 7\}$ indicating one of the 7 days (i.e., Monday, Tuesday, Wednesday, Thursday, Friday, Saturday, Sunday) in a week. The probability of a check-in at location l on day d is the probability of the location l visited on day d in the previous check-in history. For example, assume we want to predict the user's next check-in location on Saturday. According to Table 3.1, the probability of each location being visited on Saturday is shown in Table 3.7. We see that l_2 is the predicted location since it has the greatest probability of being visited on Saturday.

Table 3.7: Predicting the next location with the MFT Model (weekly patterns)

Location	l_1	l_2	l_3	l_4	l_5	l_6
No. of Visits	0	2	1	0	1	1
Check-in Probability	0	0.40	0.20	0	0.20	0.20

Improved MFT Model

In the MFT model, temporal information is discrete. The prediction of check-in location at a specific time relies on the observation of historical check-ins at that time. However, we may not have check-in data of a particular time. For example, if we want to predict the user's next check-in location at 6:00 am, we see no historical check-in in Table 3.1. Furthermore, even if there are check-ins at the given time, it would be arbitrary to consider only locations visited at this time as candidates. For example, if we want to predict the user's next check-in location at 10:30 am, although the user has check-in at l_1 and l_4 around 10:30 am before, these two locations may not be the only prediction candidates. According to the check-in time of other locations, l_2 has been mostly visited around 11:00 am. Since human movement is a stochastic process, the probability of l_2 being visited at 10:30 am should not be arbitrarily 0. Smoothing techniques are needed in order to infer the probability of a location being visited at an unobserved time.

One typical way of smoothing discrete values is to consider the values as following certain distributions. To smooth the temporal periodic patterns at different discrete time points, various distributions can be applied, as introduced in Section 2.4.5, Gaussian distribution has been observed in human mobility and proven efficient and effective for capturing temporal patterns. Gaussian distribution, also referred to as normal distribution, is a continuous probability distribution. Its probability function contains two parameters, i.e., mean μ and variance σ, as shown below:

$$f(x, \mu, \sigma) = \frac{1}{\sqrt{2\pi\sigma^2}} e^{-\frac{(x-\mu)^2}{2\sigma^2}}. \tag{3.7}$$

Figure 3.1 presents a typical normal distribution. Normal distribution has a set of properties suitable to model temporal periodic patterns; and

- probability distribution centers on a specific time point, corresponding to the mean; and

- probability decreases as the distance to the center point increases and the decreasing rate is related to the variance.

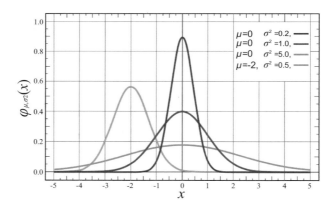

Figure 3.1: An example of Gaussian distributions with different means and variances (based on ([88])).

Human behavior exhibits similar properties regarding temporal patterns. A user visits a location during certain time period(s) (e.g., visiting a bar at night), and rarely visits that location during other time period(s) (e.g., visiting a restaurant at 3:30 pm). Figure 3.2 presents a user's check-in distribution on a specific location over 24 h of the day from a real-world mobility dataset.

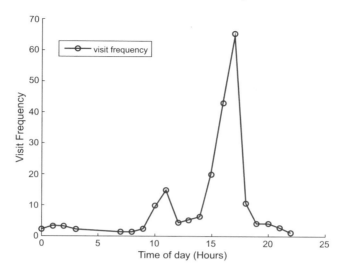

Figure 3.2: An example of a distribution from real-world mobility dataset ([30]).

It is observed that the user does not have checked-in at this location at 5:00 am. Assume we want to predict the user's check-in probability on this location at 5:00 am; the MFT model would return 0. With the improved MFT model, the check-ins in Figure 3.2 are modeled as a normal distribution over time. To predict the user's next visiting location at hour h, we follow the prediction procedure.

1. For each location l visited by the user in the observed check-in history, generate its check-in count over 24 h of the day.

2. Adapt normal distribution to model the check-in count in Step 1, compute the corresponding mean μ_l and variance σ_l.

3. Based on the μ_l and σ_l obtained in step 2, use Eq. (3.7) to compute the visiting probability $P(h|\mu_l, \sigma_l)$ with the given time h as input x, which is the probability of check-in at h for location l, i.e., $P(h|l)$.

4. Repeat Steps 1–3 to compute the probability $P(h|\mu_l, \sigma_l)$ for all the ls. Rank the probability in descending order and select the top $P(h|\mu_l, \sigma_l)$; output the corresponding l as prediction.

We use the check-ins in Figure 3.2 as an example to illustrate how it works. Let $\mathcal{H}_u = \{c_1, c_2, ..., c_N\}$ be the set of historical check-in locations of user u, where N is the total number of check-ins $\mathcal{H}_u$. Let c_i be the i-th check-in and h_i be the hour information of c_i, where $h_i \in \{0, 1, 2, ..., 23\}$.

Step 1. Generate the corresponding check-in count of l from u in $\mathcal{H}_u$ over 24 h, as shown in Table 3.8.

Table 3.8: Check-in count in $\mathcal{H}_u$ over 24 h

Hour of the Day	0	1	2	3	4	5	6	7
Count	2	3	3	2	0	0	0	1
Hour of the Day	8	9	10	11	12	13	14	15
Count	1	2	10	15	4	5	6	20
Hour of the Day	16	17	18	19	20	21	22	23
Count	43	65	11	4	4	3	1	0

Step 2. Adopt normal distribution to fit the check-in count. Consider that each check-in action c_i on l at a specific hour h_i follows the normal distribution over time as Eq. (3.7),

$$p(h_i | c_i = l, \mathcal{H}_u) \sim \mathcal{N}(h_i | \mu_l, \sigma_l^2), \tag{3.8}$$

where μ_l and σ_l^2 are the mean and variance of Gaussian distribution. Assuming the check-ins are i.i.d (independent and identically distributed), the joint probability of the training data set, given μ_l and σ_l, is in this form:

$$\prod_{i=1}^{N} p(h_i | c_i = l) \tag{3.9}$$
$$= \prod_{i=1}^{N} \mathcal{N}(h_i | \mu_l, \sigma_l^2)$$
$$= \prod_{i=1}^{N} \frac{1}{(2\pi\sigma_l^2)^{1/2}} \exp\{-\frac{1}{2\sigma_l^2}(h_i - \mu_l)^2\}.$$

By maximizing the log likelihood above, we obtain μ_l and σ_l^2,

$$\mu_l = \frac{1}{N} \sum_{i=1}^{N} h_i$$
$$\sigma_l^2 = \frac{1}{N} \sum_{i=1}^{N} (h_i - \mu_l)^2. \tag{3.10}$$

According to Figure 3.2,

$$N = 2 + 3 + 3 + 2 + 1 + 1 + 2 + 10 + 15 + 4 + 5$$
$$+ 6 + 20 + 43 + 65 + 11 + 4 + 4 + 3 + 1$$
$$= 205$$

$$\sum_{i=1}^{N} h_i = \sum_{i=1}^{N_0} h_i + \sum_{i=1}^{N_1} h_i + ... + \sum_{i=1}^{N_t} h_i + ... + \sum_{i=1}^{N_{23}} h_i$$
$$= 0 * 2 + 1 * 3 + 2 * 3 + 3 * 2 + 7 * 1 + 8 * 1 + 9 * 2 + 10 * 10$$
$$+ 11 * 15 + 12 * 4 + 13 * 5 + 14 * 6 + 15 * 20 + 16 * 43$$
$$+ 17 * 65 + 18 * 11 + 19 * 4 + 20 * 4 + 21 * 3 + 22 * 1$$
$$= 3042, \tag{3.11}$$

where N_t is the number of check-in occurrences at t.

Thus,

$$\mu_l = 3042/205 = 14.8390. \tag{3.12}$$
$$\tag{3.13}$$

Similarly,

$$\sigma_l^2 = 16.0375. \tag{3.14}$$
$$\tag{3.15}$$

Step 3. With μ_l and σ_l obtained in Step 2, compute the visiting probability at $h = 5$ for location l.

$$P(h = 5|l) = P(h = 5|\mu_l, \sigma_l) = \frac{1}{\sqrt{2\pi * 16.0375}} e^{-\frac{(5 - 14.8390)^2}{2 * 16.0375}} = 0.0049. \tag{3.16}$$

Figure 3.3 plots the estimated Gaussian distribution based on Figure 3.2. The plot shows that the distribution captures the major trend of the user's periodic mobility patterns.

Step 4. Return l corresponding to the top ranked $P(h|l)$ as prediction.

For each location l visited by this user, repeat Steps 1–3 to compute its μ_l and σ_l, and calculate its corresponding $P(h|\mu_l, \sigma_l)$ at the given hour h, indicating the probability of this check-in happening at the specified h for location l. By ranking all the $P(h|\mu_l, \sigma_l)$, we predict the location that has the highest probability of being visited at h.

Note that the above example is to give an illustration of the improved MFT model. In practice, the model is not limited to hourly patterns or normal distribution.

- By generating the check-in count according to different temporal periodic patterns in Step 1, e.g., daily or weekly patterns, the improved MFT model could make predictions in a similar fashion.

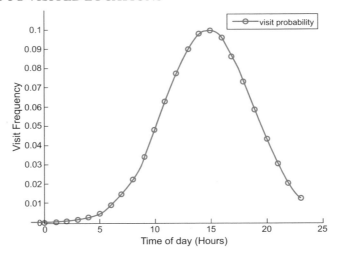

Figure 3.3: Smoothed Gaussian distribution with the Improved MFT model ([30]).

- The normal distribution can be replaced with other distributions that fit the user's check-in behavior. For example, there is work proposing the adoption of Gaussian Mixture model to model the check-in count:

$$P(h_i|c_i = l, H_u) \sim \sum_{k=1}^{2} A_k \mathcal{N}(h_i|\mu_{u,l}^k, \sigma_{u,l}^k)$$

$$\text{s.t.} \quad \sum_{k=1}^{2} A_k = 1, 0 \leq A_k \leq 1. \tag{3.17}$$

The hypothesis is that a user would like to go to the same place in two specific time periods during a day, such as visiting a restaurant for lunch and dinner, going home or working places, and so on. The check-in curve in Figure 3.2 also presents such patterns (i.e., two peaks around $h = 11$ and $h = 17$) which support this idea.

3.1.3   SOCIAL CORRELATIONS

The previous section investigates the spatio-temporal patterns for location prediction, without considering the social network information. Since LBSNs provide abundant information of both a user's spatial-temporal movements and his social networks, this section studies the role of social networks and shows how to use them in location prediction.

Social correlations [2] suggest that a user's mobile behavior is usually correlated to his social context, such as watching movies with friends or going out with colleagues. Gao et al. [28] compared the number of common check-ins between two users with friendships and those with-

Table 3.9: Numbers of check-ins shared between two users

	Average No. of Common Check-ins
Friends	11.8306
Non friends	4.3226

out friendships on a real-world LBSN dataset, as shown in Table 3.9. On average, users without friendships share approximately 4.32 check-ins, while users with friendships share approximately 11.83 check-ins, which is almost 3 times the former.

To verify the effect of social correlations in human mobility on LBSNs, i.e, whether people with friendships go to similar places than those without, many methods can be used. Here, we introduce a hypothesis testing approach. For each user, let $\mathbf{f} \in \mathbb{R}^m$ be his check-in vector with each element $\mathbf{f}(k)$ being the number of check-ins at location $l_k \in \mathcal{L}$, where $m = |\mathcal{L}|$ is the location vocabulary size. Let the similarity between u_i and a group G of other users be the average similarity between user u_j and the users in group G,

$$S_G(u_i) = \frac{\sum_{u_j \in G} sim(u_i, u_j)}{|G|}, \tag{3.18}$$

where $sim(u_i, u_j)$ denotes the similarity between u_i and u_j, which will be discussed later. For each u_i, we calculate two similarity measures, i.e., $\mathbf{S}_F(u_i)$ is the average similarity of u_i and his friendship network; $\mathbf{S}_R(u_i)$ is the average similarity of u_i and randomly chosen users, who are not in the friendship network of u_i. The number of randomly chosen users is the same as that of u_i's friends.

A two-sample t-test is conducted on the vectors $\mathbf{S}_F$ and $\mathbf{S}_R$. The null hypothesis is H_0: $\mathbf{S}_F \leq \mathbf{S}_R$, i.e., users with friendships share less common check-ins than those without, and the alternative hypothesis is H_1: $\mathbf{S}_F > \mathbf{S}_R$. It is reported in [28] that the null hypothesis is rejected at significant level $\alpha = 0.001$ with p-value of $2.6e\text{-}6$, suggesting that users with friendships have higher check-in similarity than those without.

Although several studies investigated the social correlations for next location prediction, its prediction ability has not been fully exploited. Compared to spatial and temporal information, social information is less effective in predicting a user's repetitive mobile behavior, as generally a user's repetitive mobile behavior is more affected by his personal interests than his friends' preferences. Thus, social correlation is commonly adopted as a component to assist spatial and temporal information for next location prediction. Generally, social correlation based prediction approaches investigate a user's friends' check-in history to infer the user's check-in interests to make prediction. Based on whether to consider a friend's check-in actions as i.i.d., approaches can be classified into two categories: the independent and the sequential social mobility models.

Independent Social Mobility Model

The Independent Social Mobility model [17] considers that each check-in from friends has an *independent effect* on the user's current check-in location, and models the probability of a user u's next check-in c_n on location l as a function of the geographical and temporal distance between l and each of u's friends' check-in locations.

$$P(c_n = l | t_n = t, \mathcal{H}_{F(u)}) = \sum_{(t_i, l_i) \in \mathcal{H}_{F(u)}} |t_i - t|^{-\alpha} \cdot |l_i - l|^{-\beta} \qquad (3.19)$$

where $F(u)$ is the set of u's friends. $\mathcal{H}_{F(u)}$ is the set of historical check-ins from $F(u)$. t_i and l_i are the time and location of the check-in $c_i \in \mathcal{H}_{F(u)}$. α and β are two parameters to capture the power-law property of the geographical and temporal effects.

In reality, α and β are set as empirical values, or learned from the check-in history with maximum likelihood methods.

Sequential Social Mobility Model

Different from the Independent Social Mobility model, the Sequential Social Mobility model considers that check-ins from the user's friends are correlated and their sequential patterns could affect the user's current check-in. For each friend, it considers the friend's check-in history $\mathcal{H}_{F(u)}$ as the user's check-in history, then based on the user's current check-in context $\mathcal{C}_u$ (i.e., recently visited locations), applies methods based on sequential patterns or temporal dynamics discussed above to make a prediction. Thus, for each friend, the system makes a prediction. In the end, the final prediction is made through the weighted combination on all the predictions from friends, where the weight is commonly estimated with user similarity.

Let $P(c_n = l | \mathcal{H}_{F(u)}, \mathcal{C}_u)$ be the check-in probability calculated based on u's social friends, defined as

$$P(c_n = l | \mathcal{H}_{F(u)}, \mathcal{C}_u) = \sum_{u_j \in F(u)} sim(u, u_j) P(c_n = l | \mathcal{H}_{u_j}, \mathcal{C}_u). \qquad (3.20)$$

$P(c_n = l | \mathcal{H}_{u_j}, \mathcal{C}_u)$ can be computed using Order-k Markov, Fallback Markov, and Combined Markov models.

Measuring User Similarity

User similarity on LBSNs is computed through the user's check-in history. We introduce three typical similarity measures.

- **Jaccard Index (JI)**

    The Jaccard Index, or the Jaccard similarity coefficient, is defined as the size of the intersection divided by the size of the union of the sample sets:

$$sim^{JI}(u_i, u_j) = \frac{|\mathbf{C}_i \bigcap \mathbf{C}_j|}{|\mathbf{C}_i \bigcup \mathbf{C}_j|}. \qquad (3.21)$$

Two users are more similar if their check-in location overlaps more, regardless of how frequently the check-in is.

- **Cosine Similarity (CS)**
  Two users may visit the same locations but with different visiting frequencies. For example, user u_i may visit location l_1 three times and location l_2 two times, while user u_j may visit location l_1 five times and location l_2 ten times. The Jaccard Index is unable to distinguish such check-in preference between u_i and u_j, but the Cosine similarity can. It is defined as

$$sim^{CS}(u_i, u_j) = \frac{\mathbf{C}_i \cdot \mathbf{C}_j}{|\mathbf{C}_i|_2 \times |\mathbf{C}_j|_2}. \tag{3.22}$$

where $|\mathbf{X}|_2 = \sum_k \mathbf{X}(k)$ is the 2-norm of vector $\mathbf{X}$.

- **Pearson Correlation Coefficient (PCC)**
  Different users may have different check-in preferences. Some users check-in frequently while other users rarely check-in. PCC is the measure to capture such variance.

$$sim^{PCC}(u_i, u_j) = \frac{\sum_k (\mathbf{C}_i(k) - \bar{\mathbf{C}}_i) \cdot (\mathbf{C}_j(k) - \bar{\mathbf{C}}_j)}{\sqrt{(\mathbf{C}_i(k) - \bar{\mathbf{C}}_i)^2}\sqrt{(\mathbf{C}_j(k) - \bar{\mathbf{C}}_j)^2}}, \tag{3.23}$$

where $\mathbf{C}_i(k)$ is u_i's historical check-in count on location l_k.

Example 3.5   *Give two users, u_1 and u_2, with check-ins $\{l_1, l_2, l_2, l_3, l_1, l_1, l_3, l_2, l_5\}$ and $\{l_4, l_1, l_3, l_5, l_1, l_1, l_4, l_3, l_2\}$, respectively. Compute their check-in similarity with the Jaccard Index, Cosine Similarity, and the Pearson Correlation Coefficient.*

First, we organize the historical check-ins into vectors,

$$C_1 = (3, 3, 2, 0, 1), C_2 = (3, 0, 2, 2, 1). \tag{3.24}$$

Note that the k-th element in C_i corresponds to u_i's check-in count on l_k.

$$sim^{JI}(u_i, u_j) = \frac{|\mathbf{C}_i \cap \mathbf{C}_j|}{|\mathbf{C}_i \cup \mathbf{C}_j|} = \frac{3}{5} = 0.6, \tag{3.25}$$

$$
\begin{aligned}
sim^{CS}(u_i, u_j) &= \frac{\mathbf{C}_i \cdot \mathbf{C}_j}{|\mathbf{C}_i|_2 \times |\mathbf{C}_j|_2} \\
&= \frac{3 \times 3 + 3 \times 0 + 2 \times 2 + 0 \times 2 + 1 \times 1}{\sqrt{3^2 + 3^2 + 2^2 + 1^2} + \sqrt{3^2 + 2^2 + 2^2 + 1^2}} \\
&= 0.6881,
\end{aligned} \tag{3.26}
$$

$$sim^{PCC}(u_i, u_j) = \frac{\sum_k (\mathbf{C}_i(k) - \bar{\mathbf{C}}_i) \cdot (\mathbf{C}_j(k) - \bar{\mathbf{C}}_j)}{\sqrt{(\mathbf{C}_i(k) - \bar{\mathbf{C}}_i)^2} \sqrt{(\mathbf{C}_j(k) - \bar{\mathbf{C}}_j)^2}}$$

$$= \frac{1.2 \times 1.4 + 1.2 \times (-1.6) + 0.2 \times 0.4 + (-1.8) \times 0.4 + (-0.8) \times (-0.6)}{\sqrt{1.2^2 + 1.2^2 + 0.2^2 + (-1.8)^2 + (-0.8)^2} + \sqrt{1.4^2 + (-1.6)^2 + 0.4^2 + 0.4^2 + (-0.6)^2}}$$

$$= -0.0673, \tag{3.27}$$

where $\bar{\mathbf{C}}_i = 1.8$ *and* $\bar{\mathbf{C}}_j = 1.6$.

3.1.4 HYBRID MODELS

The models discussed above focus on only one type of LBSN information, i.e., spatial sequence, temporal dynamics, or social correlations. According to available information, we can choose corresponding models to make the prediction. However, when there is more than one type of information available, different information may complement each other. For example, a user's check-in location is correlated to both his current geographical position and time. To fully exploit the heterogeneous information, a hybrid model can be considered for location prediction. Hybrid models can be classified into feature based classification models and information based probabilistic models.

Feature-Based Classification Model
The feature-based classification model applies supervised learning [39] to predict a user's check-in location. It gathers training data containing features with labels, learns a model based on the training set, and applies the model to predict the labels of the test data. We provide further details next.

1. For each observed check-in represented as a (u,l) pair, a feature vector is constructed w.r.t. the user u and location l involved in the check-in action. According to available information, features can be roughly classified as spatial features (e.g., the number of times u has previously checked-in at l, the total number of times l has been checked-in by all the users, etc.), temporal features (e.g., the number of times u has checked-in at l at different hours of the day, or days of the week, etc.), and social features (the total number of friends who checked-in at l, etc.).

   Then, a set of unobserved check-ins (u,l) are sampled. The sample size is the same as that of observed check-ins. For each pair of unobserved (u,l), a feature vector is also constructed following the same procedure for observed check-ins. The training data is organized as a set of (feature vector, label) pairs, where each observed check-in feature vector is associated with a positive label, and each unobserved check-in is associated with a negative label.

2. Classification methods, e.g., logistic regression, SVM, decision tree, can be applied on the training data to train a prediction model. Various tools can be used to learn the model, including WEKA [38] and LIBSVM [10].

3. To predict whether u would check-in at l, the learned model is applied on the feature vector constructed from the target pair (u,l).

Chang et al. [11] utilized the logistic regression model to combine a set of features extracted from Facebook data. The features include a user's previous check-ins, user's friends' check-ins, demographic data, and distance of place to user's usual location. Their results demonstrate that the number of previous check-ins by the user is a strong predictor, while previous check-ins made by friends and the age of the user are also good features for prediction. The regression model is formulated as

$$P(c_n = l | t_n = t, X_t^u) = \frac{1}{1 + exp^{-X_t^u \theta}}, \tag{3.28}$$

where $X_t^u \in \mathbb{R}$ is a vector of features of user u at time t, and $\theta \in \mathbb{R}$ is a vector of feature parameters.

Information Based Probabilistic Model

The information based probabilistic model first makes predictions with each type of information, and makes the final prediction by combining all predictions through certain probabilistic assumptions. For example, linear combination is mostly used for integrating social friendship with spatio-temporal patterns. The probability of a user u's next visit at a location l, i.e., $P(c_n = l)$, is computed as a weighted combination of both social effect $P(c_n = l | \mathcal{H}_{F(u)})$ and non-social effect $P(c_n = l | \mathcal{H}_u)$, as shown in Eq. (3.29). The probability of social effect is based on the user's friends' check-in history $\mathcal{H}_{F(u)}$, while the probability of non-social effect considers the user's own check-in history $\mathcal{H}_u$:

$$P(c_n = l) = \alpha P(c_n = l | \mathcal{H}_u) + (1 - \alpha) P(c_n = l | \mathcal{H}_{F(u)}). \tag{3.29}$$

Cho et al. [17] considered the user check-in probability as a linear combination of social effect and non-social effect. The social effect is modeled through the Independent Social Mobility Model, while the non-social effect is about the periodic patterns, which consider the user's personal movement following a 2-D Gaussian distribution, with the two Gaussian centers focusing on home and work.

Gao et al. [28] proposed a social-historical model integrating the social ties and historical ties of a user for location prediction, as illustrated in Figure 3.4. Both ties generate the probability of the next location based on the observation of the previous check-in sequence with the Combined Markov model. The historical ties consider the user's own check-in sequence, and the social ties consider the check-in sequences of the user's friends. A parameter $\eta \in [0, 1]$ is introduced to control the weight between historical ties and social ties. For a particular user u_i, the probability of the next check-in location is defined as

$$P_{SH}^i(c_n = l) = \eta P_H^i(c_n = l) + (1 - \eta) P_S^i(c_n = l), \tag{3.30}$$

where $P_H^i(c_n = l)$ is the probability of u_i's check-in at location l from his historical ties, and $P_S^i(c_n = l)$ from his social ties. Social information makes around a 20–30% contribution in predicting a user's next location.

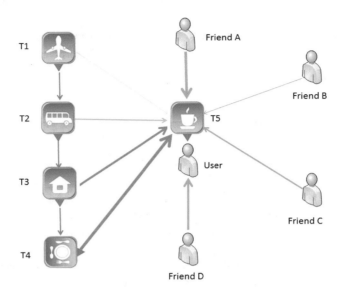

Figure 3.4: Location prediction with social-historical ties ([28]).

Conditional probability is also introduced to integrate the spatial information and temporal information. Given a series of historical visits in a previous time section $\mathcal{H}_u$, and a context of the latest visit location $\mathcal{C}_u$ with the time of the next visit t_n, the location prediction problem can be described as finding the probability

$$P(c_n = l | t_n = t, \mathcal{H}_u, \mathcal{C}_u), \tag{3.31}$$

Using Bayes' rule, the probability in Eq. (3.31) is equivalent to:

$$
\begin{aligned}
&P(c_n = l | t_n = t, \mathcal{H}_u, \mathcal{C}_u) \\
&= \frac{P(c_n = l, t_n = t | \mathcal{H}_u, \mathcal{C}_u)}{P(t_n = t)} \\
&\propto P(c_n = l, t_n = t | \mathcal{H}_u, \mathcal{C}_u) \\
&= P(t_n = t | c_n = l, \mathcal{H}_u, \mathcal{C}_u) P(c_n = l | \mathcal{H}_u, \mathcal{C}_u) \\
&= P(t_n = t | c_n = l, \mathcal{H}_u) P(c_n = l | \mathcal{H}_u, \mathcal{C}_u).
\end{aligned}
\tag{3.32}
$$

Note that we consider

$$P(t_n = t | c_n = l, \mathcal{H}_u, \mathcal{C}_u) = P(t_n = t | c_n = l, \mathcal{H}_u) \tag{3.33}$$

under the assumption that the probability of the current visit time is relevant to the current visit location but not other recently visited locations.

As we can see, the first part $P(t_n = t | c_n = l, \mathcal{H}_u)$ in Eq. (3.32) can be computed with MFT-based model based on temporal dynamics, and the second part $P(c_n = l | \mathcal{H}_u, C_u)$ can be computed with Markov-based model based on sequential patterns.

Most current works report very limited improvement by utilizing social network information in LBSNs. The model that considers social networks performs slightly better than that does not consider social networks. However, we cannot conclude that social networks have no contributions to a user's mobility. It is still an open problem.

3.2 HOME LOCATION PREDICTION

The motivation of home location prediction arises from the sparsity of available user home locations on popular social networks such as Twitter and Facebook. Based on the statistics from [15], only 26% of Twitter users list their locations with granularly as a city name, and fewer than 0.42% of all tweets use the geo-tagging function to indicate their locations. The awareness of user home location could provide an opportunity to study geo-social networks from a user's egocentric view, and improve targeting advertisement regions and better summarize the local news for nearby users. Therefore, obtaining the user home location is helpful to studying human mobility on location-based social networks.

In social media, a user commonly specifies his hometown information in his profile at a city level, which limits the geographical granularity of the ground truth we can generate. Thus, the prediction of home location is commonly also at a city level, which could be compared with the ground truth to compute the prediction accuracy.

The home location prediction task becomes trivial once sufficient check-in history can be observed from a user, as one intuitive way is to find the city which contains the majority of the user's historical check-ins as the prediction. Improvements can be made by taking the geographical average on the user's historical check-in locations within the city, which may output a home location with geographical granularity finer than the city level. However, given the geographical sparsity discussed above, a large number of users do not have sufficient check-in history, and the home location prediction task is specifically proposed for these users. Many other types of information can be used for prediction.

Among current approaches, network information and content information are the two most widely used types of information. The content-based approaches [15, 40] study the location information implicated in a user's tweet content, and they make location prediction based on the correlation between specific terms in tweets and their corresponding locations.

3.2.1 CONTENT-BASED PREDICTION

Content-based home location prediction uses the content published by a user to predict the user's home location. Let W_u denote the user u's content in terms of words and l_u denote the user's location. Generally, the content-based methods aim to model the relationship between W_u and u for prediction. Given the words published by a user, the following probabilistic model is investigated by Cheng et al. [15]:

$$p(l|\mathbf{w}_u) = \sum_{w \in \mathbf{w}_u} p(l|w) * p(w), \tag{3.34}$$

where w is a word in W_u, and the location probability is computed as the aggregation of all words posted by the user. $p(w)$ and $p(l|w)$ are pre-calculated through the training data. $p(w)$ is defined as the prior probability of w in the whole corpus,

$$p(w) = \frac{count(w)}{N}, \tag{3.35}$$

where $count(w)$ is the number of occurrences of the word w, and N is the total number of tokens in the corpus. $p(l|w)$ is the probability of the location being city l given the observed word w,

$$p(l|w) = \frac{count(w,l)}{count(w)}, \tag{3.36}$$

where $count(w,l)$ is the number of occurrence of w in city l. Laplace smoothing is commonly applied to avoid the 0 value,

$$p(l|w) = \frac{count(w,l) + 1}{count(w) + V}, \tag{3.37}$$

where V is the total number of cities.

The above approach directly models the relationship between the content and location. Another method assumes that the user's home locations can be represented as a certain function. For example, Eisenstein et al. [22] adopted a linear function to model the relationship

$$\begin{aligned} l_u^{lat} &= \mathbf{w}_u^T \mathbf{a}^{lat} \\ l_u^{lng} &= \mathbf{w}_u^T \mathbf{a}^{lng}, \end{aligned} \tag{3.38}$$

where l^{lat} and l^{lng} are latitude and longitude of location l, $\mathbf{w}_u$ is the word vector extracted from the user's published content, and $\mathbf{a}^{lat}$ and $\mathbf{a}^{lng}$ are the corresponding coefficients. The prediction problem is then modeled as the following minimization problem

$$\begin{aligned} \sum_u (\mathbf{w}_u^T \mathbf{a}^{lat} - l_u^{lat})^2 + (\mathbf{w}_u^T \mathbf{a}^{lng} - l_u^{lng})^2 \\ + \lambda_{lat} \|\mathbf{a}^{lat}\|_1 + \lambda_{lng} \|\mathbf{a}^{lng}\|_1, \end{aligned} \tag{3.39}$$

where l^{lat}, l^{lng}, and $\mathbf{w}_u$ are extracted from (content, location) pairs in the training data. The learning process of this model is beyond the scope of this book. For more details, the reader can refer to [22]. By solving the above optimization problem, $\mathbf{a}^{lat}$ and $\mathbf{a}^{lng}$ can be learned. Then, for a new user, via obtaining his content organized as $\mathbf{w}_u$, the predicted location of the user is outputted as $(\hat{l}_u^{lat}, \hat{l}_u^{lng})$ with each element computed through Eq. (3.38).

3.2.2   NETWORK-BASED PREDICTION

Different from content-based prediction, network-based prediction focuses on the relationship between geographical distance and social friendship, referred to as the inverse distance rule discussed in Section 2.4.1. Specifically, it infers a user's home location from his social friends' home locations.

Backstrom et al. [3] pioneered work in home location prediction with social friendships on Facebook. They discover that the probability of a link being present between two nodes is a function of their geographical distance according to Figure 3.5,

$$p(x) = a(b + x)^{-c}, \tag{3.40}$$

where the parameters a, b, and c are empirically determined. The exponent in this function is close to -1, which indicates that the probability of friendship between two users is roughly inversely proportional to their geographical distance.

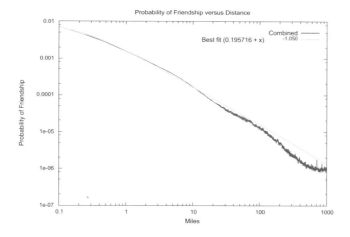

Figure 3.5: The probability of friendship over geographical distance ([3]).

With the parameters a, b, and c available, one can compute the likelihood of a user u's social connections if he lives at a location l_u based on the observed home locations of his friends. Let L_u^f be the set of locations of u's friends; the likelihood for all the connections of u whose home location is l_u can be modeled as

$$\prod_{l_f \in L_u^f} p(|l_u - l_f|) \prod_{l_r \notin L_u^f} 1 - p(|l_u - l_r|), \tag{3.41}$$

where $(l_u - l_f)$ denotes the distance between u and f, and $p(|l_u - l_f|)$ is the probability of observed friendship connection between u and f. $1 - p(|l_u - l_r|)$ is the probability of non-observed connection between u and r.

Since the computing of $\prod_{l_r \notin L_u^f} 1 - p(|l_u - l_r|)$ is a bit expensive, an alternative approach is found. Let L be the locations of all the users; the likelihood can be defined as

$$\prod_{l_f \in L_u^f} \frac{p(|l_u - l_f|)}{1 - p(|l_u - l_f|)} S_u, \tag{3.42}$$

where $S_u = \prod_{l_r \in L} 1 - p(|l_u - l_r|)$ is computed between each user u and every other user r, which is independent to u's social connections. Thus, for each u, his S_u can be pre-computed. To predict the home location of user u, we find a location l_u which maximize the likelihood of Eq. (3.42). Since l_u is mostly likely co-located with one of u's friends, we evaluate the likelihood of each location of u's friends, and pick the one with the highest probability as home location.

McGee et al. [68] further improved this approach by considering the tie strength of users' social connections. The basic idea is that instead of using a unique function in Eq. (3.42) to model all the social connections, connections with different tie-strengths could be treated differently. Thus, it firstly classifies the connections into different groups according to their different tie-strengths. Then, for each group of connections G_i, its corresponding function p_{G_i} in Eq. (3.40) w.r.t. the parameter set $(a_{G_i}, b_{G_i}, c_{G_i})$ is determined. The final likelihood is the aggregation of all the connections with their corresponding p_{G_i}.

According to McGee et al. [68], the term S_u has slight influence in the final prediction performance. Thus, if the computation time is a concern in performance, S_u can be removed from Eq. (3.42) without much performance loss.

3.3   EVALUATION METRICS

There are two commonly used evaluation metrics for location prediction, *accuracy* and *distance*. The former measures the prediction accuracy, the fraction of correctly predicted locations over the total number of predicted locations in the testing set. It is widely used in the next-location prediction task [3, 17, 28]. Sometimes its variants are used. For example, top-k accuracy is utilized in [15]. It returns the top k candidates of prediction for a location, and treats a prediction as correct as long as the correct location is among the top k returned locations. Here, k is usually selected as 2, 3, 5, and 10.

For home location prediction, *distance* is used to evaluate the performance. *Expected Distance Error* [15, 17] computes the geographical distance between the real location and the estimated location, averaged over all predicted locations.

$$AvgErrD(U) = \frac{1}{|U|} \sum_{u \in U} Err(u), \tag{3.43}$$

where U is the set of users whose locations to be predicted in the test data. $Err(u)$ is defined as the distance error of user u,

$$Err(u) = d(l_{act}(u), l_{est}(u)), \tag{3.44}$$

where $l_{act}(u)$ is the actual location of u, $l_{est}(u)$ is the estimated location of u, and $d(x, y)$ is a function that computes the geographical distance between two locations x and y.

Similarly, *accuracy@K* of distance error is also used to evaluate the performance of the home location prediction [68].

$$ACC(U)@K = \frac{|\{u \in U : Err(u) \leq K\}|}{|U|},$$

(3.45)

where K is a constant commonly set as 25 (miles).

3.4   SUMMARY

The task of location prediction focuses on capturing a user's repetitive check-in behavior and predicting which visited location the user would like to visit again. According to the power-law distribution of check-ins, both repetitive check-ins and cold-start check-ins are important in human mobility. Both academia and industry pay attention to location prediction and develop state-of-the-art methods for capturing human repetitive check-in behavior for location prediction.

The study of repetitive check-in behavior relies on the analysis of a user's check-in history. Location prediction methods generally leverage the historical information of a user's check-in behavior and his network information to predict his locations, including next visited locations and home locations. Algorithms for next visited location prediction can be classified into sequential patterns, temporal dynamics, social correlations, and hybrid models. Among them, social correlations can help predict not only a visited location, but also a new location. Since social correlations only correspond to 30% prediction effect, hybrid models considering social correlations and spatio-temporal patterns usually output a visited location.

In the next chapter, location recommendation methods are introduced to study check-ins on new locations. Spatial, temporal, social, and content-based models focusing on this behavior will be discussed for new location prediction.

CHAPTER 4

Finding New Locations to Visit

An increasing number of locations with specific functions are called points of interest (POIs), e.g., restaurants, theaters, stores, and hotels, to enrich people's life and entertainment. Generally, people want to explore the city and neighborhood in their daily life, and find a "new" place to go (e.g., a new restaurant, a new store) which they have never visited before. To predict where to go is a decision-making problem related to the user's personal interests. A large number of POIs presents a problem of "choice paralysis" [8]. When exploring a new place, we are facing too many location choices and do not know which one best matches our interests. This corresponds to a task called *location recommendation*, which aims to help users filter out uninteresting POIs and find a satisfying decision [105, 118].

In the last decade, recommender systems have been widely studied among various categories, such as movie recommendation [48], job recommendation [99], item recommendation [90], and news recommendation [92]. Location recommendation is a sub-category of recommender systems. Thus, technologies that applicable to general recommender systems can also be considered for location recommendation with some performance loss [91]. In the following sections, we first introduce the general recommender systems, and then discuss how to design location recommender systems based on mobility patterns on LBSNs.

4.1 RECOMMENDER SYSTEMS

Recommender systems refers to platforms that apply knowledge discovery techniques to analyze user preferences and make recommendations about information, items, or services. The techniques used by recommender systems can be generally classified into three main categories: collaborative filtering, content-based, and hybrid recommendation, as shown in Figure 4.1. Among them, collaborative filtering is the most successful, which has been proven effective and efficient in practice [89].

Collaborative Filtering (CF) considers that two users with similar behavior in the past (e.g., watching similar movies, purchasing similar products, visiting similar restaurants) would have similar behavior in the future. It contains memory-based and model-based approaches. Before discussing each approach, we first introduce the data representation of check-ins for collaborative filtering.

4.1.1 CHECK-IN DATA REPRESENTATION

Generally, check-ins are organized as a user-location matrix, as shown in Table 4.1.

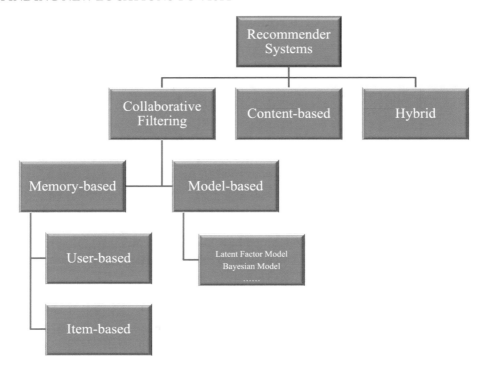

Figure 4.1: Hierarchy of general recommender systems.

Table 4.1: An example of user-location matrix

	l_1	l_2	l_3	l_4	l_5	l_6	l_7
u_1	1		3	2		4	
u_2	2	3		3		2	4
u_3	3		2		3		5
u_4			2		8		2
u_5		4	7			2	3

The above check-in matrix is denoted as $\mathbf{C}$ with each entry $\mathbf{C}_{i,j}$ indicating that u_i has checked-in at l_j for $\mathbf{C}_{i,j}$ times. The empty entry denotes that there is no check-in. For example, $C_{2,4} = 3$ indicates that u_2 has visited l_4 three times; $\mathbf{C}_{2,3}$ and $\mathbf{C}_{2,5}$ are empty, indicating that u_2 has not checked-in at l_3 and l_5. Note that sequential information of check-ins is lost in matrix representation.

A user can check-in at a location many times, e.g., visiting a favorite restaurant every week. This could make certain entry values in the matrix much greater than others. Such entries may skew the computation of user preference similarity and affect the recommendation quality. Thus, two alternatives have been widely used to replace the original check-in matrix. One is binary representation, and the other is normalized representation.

In the binary representation, each entry is represented by 0 or 1 based on whether a checked-in is observed. If a user has checked-in at a location, the corresponding entry is filled with 1, otherwise 0 or left empty. Table 4.2 presents the binary representation of check-ins from Table 4.1.

Table 4.2: The binary representation of check-in actions

	l_1	l_2	l_3	l_4	l_5	l_6	l_7
u_1	1	0	1	1	0	1	0
u_2	1	1	0	1	0	1	1
u_3	1	0	1	0	1	0	1
u_4	0	0	1	0	1	0	1
u_5	0	1	1	0	0	1	1

In the normalized representation, each entry value is mapped to [0,1] with a pre-defined normalization function. Table 4.3 lists some common functions. For example, using an inverse function to normalize the check-in matrix, we obtain the representation in Table 4.4.

Table 4.3: Normalization functions for check-in matrix

Normalization Strategy	Row-based	Column-based	Matrix-based
sum	$\hat{C}_{i,j} = \frac{C_{i,j}}{\sum_j C_{i,j}}$	$\hat{C}_{i,j} = \frac{C_{i,j}}{\sum_i C_{i,j}}$	$\hat{C}_{i,j} = \frac{C_{i,j}}{\sum_i \sum_j C_{i,j}}$
max	$\hat{C}_{i,j} = \frac{C_{i,j}}{\max_j (C_{i,j})}$	$\hat{C}_{i,j} = \frac{C_{i,j}}{\max_i (C_{i,j})}$	$\hat{C}_{i,j} = \frac{C_{i,j}}{\max_{i,j} (C_{i,j})}$
sigmod function		$\hat{C}_{i,j} = \frac{1}{1 + e^{-C_{i,j}}}$	
inverse function		$\hat{C}_{i,j} = \frac{1}{1 + C_{i,j}^{-1}}$	

The choice of normalization functions depends on data and recommendation algorithms. Generally, the normalized representation works better than the binary representation since the latter does not distinguish a user's preferences on different visited locations.

4.1.2 MEMORY-BASED COLLABORATIVE FILTERING

The memory-based collaborative filtering uses entire check-in matrix to recommend items. It has been adopted in many commercial systems. According to whose similarity it relies on to perform

Table 4.4: Normalizing check-in actions with the inverse function

	l_1	l_2	l_3	l_4	l_5	l_6	l_7
u_1	0.5000		0.7500	0.6667		0.8000	
u_2	0.6667	0.7500		0.7500		0.6667	0.8000
u_3	0.7500		0.6667		0.7500		0.8333
u_4			0.6667		0.8889		0.6667
u_5		0.8000	0.8750			0.6667	0.7500

the recommendation, approaches can be classified into user-based and item-based collaborative filtering.

User-based Collaborative Filtering

The idea of user-based collaborative filtering is to capture a user u's preferences on unvisited locations based on the K users most similar to him. It contains three steps.

1. Select K users most similar to u as his neighborhood $\mathcal{N}_u$. User similarity is computed based on a similarity measure, as discussed in Chapter 3.

2. Aggregate the preferences of users from $\mathcal{N}(u)$ on the locations unvisited by u, and deem them as u's preferences. Equation (4.1) is a widely used aggregation function.

$$\hat{C}_{i,j} = \bar{C}_i + \frac{\sum_{u_k \in \mathcal{N}(u_i)} sim(u_i, u_k)(C_{k,j} - \bar{C}_k)}{\sum_{u_k \in \mathcal{N}(u_i)} sim(u_i, u_k)}, \qquad (4.1)$$

where $\bar{C}_i$ is the average check-in count of u_i on his visited locations, defined as

$$\bar{C}_i = \frac{1}{|I_i|} \sum_{j \in I_i} C_{i,j}, \qquad (4.2)$$

where I_i represents u_i's visited locations.

3. Rank u's preferences on those unvisited locations and select the top N locations for recommendation.

Below is an illustrative example of user-based collaborative filtering.

Example 4.1 *Given the check-in matrix in Table 4.1, recommend a location to u_1 with user-based collaborative filtering.*

In the first step, vector cosine similarity between u_1 to other users is computed based on their check-in count. For example, the similarity between u_1 and u_3 is

$$sim(u_1, u_3) = \frac{1 * 4 + 3 * 6 + 2 * 0 + 0 * 1}{\sqrt{(1^2 + 3^2 + 2^2)}\sqrt{(4^2 + 6^2 + 1^2)}} = 0.8076. \tag{4.3}$$

Table 4.5 lists the similarities related to u_1.

Table 4.5: Check-in similarity between u_1 and other users

	u_2	u_3	u_4	u_5
u_1	0.4507	0.2397	0.1291	0.5995

We set $K = 3$, and select the top K users most similar to u_1, corresponding to the neighborhood set $\mathcal{N}(u_1) = \{u_2, u_3, u_5\}$.

In the second step, we compute the average check-in count of each user, listed in Table 4.6, and then compute the preferences of u_1 on his unvisited locations $\{l_2, l_5, l_7\}$ as in Eq. (4.1).

Table 4.6: Average check-in count of each user

	u_1	u_2	u_3	u_4	u_5
Avg. Check-in Count	2.5	2.8	3.25	4	4

$$\hat{C}_{1,2} = 2.5 + \frac{0.4507 * (3 - 2.8) + 0.5995 * (4 - 4)}{0.4307 + 0.5995} = 2.5875$$

$$\hat{C}_{1,5} = 2.5 + \frac{0.2397 * (3 - 3.25)}{0.2397} = 2.25$$

$$\hat{C}_{1,7} = 2.5 + \frac{0.4507 * (4 - 2.8) + 0.2397 * (5 - 3.25) + 0.5995 * (3 - 4)}{0.4307 + 0.2397 + 0.5995} = 2.7841 \tag{4.4}$$

In the third step, the ranking list of $u_1's$ preferences on l_2, l_5, and l_7 is generated, i.e., $(l_7 > l_2 > l_5)$. Assuming that $N = 1$ for top N recommendation, the user-based collaborative filtering returns l_7 as the recommended location to u_1.

Item-based Collaborative Filtering

Item-based collaborative filtering adapts a similar recommendation strategy as user-based collaborative filtering. Instead of finding similar users, it seeks similar locations and aggregates the check-in counts of these locations with Eq. (4.5). It computes $\hat{C}_{i,j}$ for each unvisited l_j by u_i in order to perform recommendation.

$$\hat{C_{i,j}} = \bar{C}_j + \frac{\sum_{l_k \in \mathcal{N}(l_j)} sim(l_j, l_k)(C_{i,k} - \bar{C}_k)}{\sum_{l_k \in \mathcal{N}(l_j)} sim(l_j, l_k)}, \tag{4.5}$$

where $\bar{C}_j$ is the average check-in count of l_j, $sim(l_j, l_k)$ is the check-in similarity between l_j and l_k, $\mathcal{N}(l_j)$ contains K locations most similar to l_j.

Item-based collaborative filtering can be used in another perspective of recommendation, i.e., targeting users who may be interested in a given location. For example, a restaurant prepares 100 coupons of its cuisine and plans to distribute them to new customers. In order to make the best use of these coupons, the restaurant wants to seek 100 customers who are potentially more interested in the cuisine than other users. The item-based collaborative filtering can help on this purpose. The following example illustrates its recommendation process.

Example 4.2   *Given the check-in matrix in Table 4.1, recommend a set of users who may be interested in l_2 with item-based collaborative filtering.*

Fist, cosine similarity between l_2 and other locations is computed based on their check-in count. Table 4.7 lists the similarities related to l_2. With $K = 3$, the top K locations most similar to l_2 are selected, resulting in the neighborhood set $\mathcal{N}(l_2) = \{l_3, l_6, l_7\}$.

Table 4.7: Check-in similarity between l_2 and other locations

	l_1	l_3	l_4	l_5	l_6	l_7
l_2	0.3207	0.6893	0.4992	0	0.5715	0.6532

Second, we compute the preferences of users who have not visited l_2 as in Eq. (4.5):

$$\hat{C_{1,2}} = 3.5 + \frac{0.6893 * (3 - 3.5) + 0.5715 * (4 - 2.67)}{0.6893 + 0.5715} = 3.8295$$

$$\hat{C_{3,2}} = 3.5 + \frac{0.6893 * (2 - 3.5) + 0.6532 * (5 - 3.5)}{0.6893 + 0.6532} = 4.0746$$

$$\hat{C_{4,2}} = 3.5 + \frac{0.6893 * (2 - 3.5) + 0.6532 * (2 - 3.5)}{0.6893 + 0.6532} = 1.2318$$

$$\tag{4.6}$$

For top-1 recommendation, u_3 is recommended with the highest recommendation value on l_2.

4.1.3   MODEL-BASED COLLABORATIVE FILTERING

Memory-based collaborative filtering is efficient and easy to adopt. However, there are two shortfalls when it is applied to large-scale and sparse data.

- **Sparsity**

  LBSN data is sparse due to the user-driven check-in property. The density of a check-in matrix is commonly between 10^{-4} and 10^{-5}. With sparse data, the similarity measured from check-ins could be unreliable [74]. In an extreme case, say the "cold-start" problem, a new user with no check-in history has the similarity of 0 to any other users.

- **Scalability**

  Memory-based collaborative filtering makes use of the whole check-in matrix to perform recommendation, which requires a large storage space. In addition, the computation of K nearest neighbors is inefficient with a large number of users or items.

Model-based collaborative filtering is proposed to address these issues. It uses data mining and machine learning techniques to learn a model from training data, and applies the model to test data to predict users' preferences on different locations. Typical approaches in this category include Bayesian Models, Latent Factor Models, and Classification/Regression Models. Among them, latent factor models such as matrix factorization have been widely used.

Matrix Factorization for Recommendation

The basic idea of matrix factorization is to assume that there are certain latent factors related to both users' interests and locations' properties. As an example of restaurant check-ins, latent factors could be the taste, quality, environment, price, etc. A check-in action on the restaurant is resulted by the combinational effects of a user's interests and the restaurant's properties on these factors. For example, if a user would like to have seafood in a romantic environment, he may be interested in a restaurant that serves fresh seafood with beautiful ocean view.

Let $\mathbf{u} = \{u_1, u_2, ..., u_m\}$ be the set of users, and $\mathbf{l} = \{l_1, l_2, ..., l_n\}$ the set of locations, where m and n denote the numbers of users and locations, respectively. Let $\mathbf{U} \in \mathbb{R}^{m \times d}$ be the user check-in interests and $\mathbf{L} \in \mathbb{R}^{n \times d}$ the location properties, both on latent factors, with $d \ll min(m,n)$ being the number of latent factors. With matrix factorization, each check-in actions $C_{i,j}$ is approximated as a product of two vectors $\mathbf{U}_i$ and $\mathbf{L}_j$,

$$\mathbf{C}_{ij} \approx \mathbf{U}_i \mathbf{L}_j^{\top}, \qquad (4.7)$$

where $\mathbf{C} \in \mathbb{R}^{m \times n}$ is the check-in matrix as in Table 4.1.

Location recommendation models with matrix factorization approximate u_i's check-in preference on an unvisited l_j by solving the following optimization problem:

$$\min_{\mathbf{U},\mathbf{L}} \sum_{i}^{m} \sum_{j}^{n} (\mathbf{C}_{ij} - \mathbf{U}_i \mathbf{L}_j^{\top})^2. \qquad (4.8)$$

In this model, each observed check-in $\mathbf{C}_{i,j}$ is approximated via $\mathbf{U}_i \mathbf{L}_j^{\top}$. Their difference, $\mathbf{C}_{i,j} - \mathbf{U}_i \mathbf{L}_j^{\top}$, corresponds to the approximation error. The goal is to minimize this error and make $\mathbf{U}_i \mathbf{L}_j^\top$ as close to $\mathbf{C}_{i,j}$ as possible. The outputs are user latent interests $\mathbf{U}$ and location latent properties $\mathbf{L}$.

After obtaining $\mathbf{U}$ and $\mathbf{L}$, the missing value in $\mathbf{C}$, $\widetilde{\mathbf{C}}_{ij}$, indicating the preference of u_i on an unvisited location l_j, is then approximated by $\mathbf{U}_i \mathbf{L}_j^\top$. To avoid over-fitting, two regularization terms are applied to $\mathbf{U}$ and $\mathbf{L}$, respectively. Equation (4.8) can be rewritten as

$$\min_{\mathbf{U},\mathbf{L}} \|\mathbf{C} - \mathbf{U}\mathbf{L}^\top\|_F^2 + \alpha\|\mathbf{U}\|_F^2 + \beta\|\mathbf{L}\|_F^2, \tag{4.9}$$

where α and β are non-negative parameters to control the capability of $\mathbf{U}$ and $\mathbf{L}$ for avoiding over-fitting. $\|\cdot\|_F$ is the Frobenius norm of a matrix and $\|\mathbf{X}\|_F = \sqrt{\sum_i^m \sum_j^n \mathbf{X}_{i,j}^2}$.

Since there are multiple variables in Eq. (4.9), the alternating algorithm is commonly applied to find optimal solutions for $\mathbf{U}$ and $\mathbf{L}$. The key idea is to minimize Eq. (4.9) w.r.t. one variable while fixing the other variable(s). It works as follows.

1. Initialize $\mathbf{L}$ randomly.

2. Fix $\mathbf{L}$; update $\mathbf{U}$ by minimizing Eq. (4.9).

3. Fix $\mathbf{U}$; update $\mathbf{L}$ by minimizing Eq. (4.9).

4. Repeat Steps 2 and 3 until $\mathbf{U}$ and $\mathbf{L}$ converge or reach a predefined number of maximum iterations.

Steps 2 and 3 are the most important. Several algorithms can be used to solve the minimization problem. We introduce a typical one, **Gradient Descent**.

The **Gradient Descent** algorithm is used to find a local minimum of a function. It starts at an initial point, and moves towards the local minimum through a set of steps. Each step relates to two factors: (1) step direction; and (2) step size. The step direction is computed as the negative of the gradient of the function at the current point. The step size is commonly set as empirical values or chosen through Goldstein Conditions [47]. For ease of presentation, we choose a fixed value in the following example. Generally, given an optimization function

$$\min_x f(x). \tag{4.10}$$

Gradient Descent first initializes x with a random value x_0, and then updates x iteratively through the following rule:

$$x_1 = x_0 - \delta * \frac{\partial f}{\partial x}|_{x=x_0}$$
$$x_0 = x_1, \tag{4.11}$$

where $-\frac{\partial f}{\partial x}|_{x=x_0}$ is the step direction at x_0, computed as the negative of the partial derivation of f at x_0. δ is the corresponding step size. The updating process iterates until x converges.

We apply the Gradient Descent algorithm to solve Equation (4.9). According to the matrix trace operation,

$$\|\mathbf{X}\|_F^2 = Tr(\mathbf{X}^\top \mathbf{X}), \tag{4.12}$$

where $Tr(\mathbf{X}) = \sum_i \mathbf{X}_{i,i}$ is the matrix trace, which sums all the diagonal values of $\mathbf{X}$.

Thus, Eq. (4.9) is equivalent to

$$\min_{\mathbf{U},\mathbf{L}} Tr\left((\mathbf{C} - \mathbf{U}\mathbf{L}^\top)^\top (\mathbf{C} - \mathbf{U}\mathbf{L}^\top)\right) + \alpha Tr(\mathbf{U}^\top \mathbf{U}) + \beta Tr(\mathbf{L}^\top \mathbf{L}). \tag{4.13}$$

Expanding Eq. (4.13), we obtain,

$$\min_{\mathbf{U},\mathbf{L}} Tr(\mathbf{C}^\top \mathbf{C} - \mathbf{C}^\top \mathbf{U}\mathbf{L}^\top - \mathbf{L}\mathbf{U}^\top \mathbf{C} + \mathbf{L}\mathbf{U}^\top \mathbf{U}\mathbf{L}^\top) + \alpha Tr(\mathbf{U}^\top \mathbf{U}) + \beta Tr(\mathbf{L}^\top \mathbf{L}). \tag{4.14}$$

Equation (4.14) is commonly referred to as a objective function, denoted as $\mathcal{J}$. Taking the partial derivation of $\mathcal{J}$ on $\mathbf{U}$ and $\mathbf{V}$,

$$\frac{\partial \mathcal{J}}{\partial \mathbf{U}} = -2\mathbf{C}\mathbf{L} + 2\mathbf{U}\mathbf{L}^\top \mathbf{L} + 2\alpha \mathbf{U} \tag{4.15}$$

$$\frac{\partial \mathcal{J}}{\partial \mathbf{L}} = -2\mathbf{C}^\top \mathbf{U} + 2\mathbf{L}\mathbf{U}^\top \mathbf{U} + 2\beta \mathbf{L}. \tag{4.16}$$

With the above derivations, Gradient Descent solves Eq. (4.9) as follows.

1. Initialize $\mathbf{U}$ and $\mathbf{L}$ randomly.

2. Compute $\frac{\partial \mathcal{J}}{\partial \mathbf{U}}$; update $\mathbf{U} \leftarrow \mathbf{U} - \delta_\mathbf{U} * \frac{\partial \mathcal{J}}{\partial \mathbf{U}}$.

3. Compute $\frac{\partial \mathcal{J}}{\partial \mathbf{L}}$; update $\mathbf{L} \leftarrow \mathbf{L} - \delta_\mathbf{L} * \frac{\partial \mathcal{J}}{\partial \mathbf{L}}$.

4. Repeat Steps 2 and 3 until $\mathbf{U}$ and $\mathbf{L}$ converge or reach a predefined number, $iter_{max}$, of maximum iterations.

Example 4.3 *Given the check-in matrix in Table 4.1, solve the matrix factorization problem in Eq. (4.9) with Gradient Descent.*

Step 1. Set up the input variables $\{\mathbf{U}_0, \mathbf{L}_0, \alpha, \beta, iter_{max}, d\}$. We set α and β as 0.1, $iter_{max}$ as 1000, the number of latent factors d as 2, and randomly initialize U and L as in Eq. (4.17):

$$\mathbf{U} = \begin{pmatrix} 0.2581 & 0.7112 \\ 0.4087 & 0.2217 \\ 0.5949 & 0.1174 \\ 0.2622 & 0.2967 \\ 0.6028 & 0.3188 \end{pmatrix} \quad \mathbf{L} = \begin{pmatrix} 0.4242 & 0.7303 \\ 0.5079 & 0.4886 \\ 0.0855 & 0.5785 \\ 0.2625 & 0.2373 \\ 0.8010 & 0.4588 \\ 0.0292 & 0.9631 \\ 0.9289 & 0.5468 \end{pmatrix}. \tag{4.17}$$

Step 2. Update $\mathbf{U}$ and $\mathbf{V}$ iteratively. For simplicity, we empirically set both δ_U and δ_L as 0.01. In the first iteration, compute $\frac{\partial \mathcal{J}}{\partial \mathbf{U}}$:

$$\frac{\partial \mathcal{J}}{\partial \mathbf{U}} = -2\mathbf{CL} + 2\mathbf{UL}^\top \mathbf{L} + 2\alpha \mathbf{U}$$

$$= -2 \begin{pmatrix} 1 & 0 & 3 & 2 & 0 & 4 & 0 \\ 2 & 3 & 0 & 3 & 0 & 2 & 4 \\ 3 & 0 & 2 & 0 & 3 & 0 & 5 \\ 0 & 0 & 2 & 0 & 8 & 0 & 2 \\ 0 & 4 & 7 & 0 & 0 & 2 & 3 \end{pmatrix} * \begin{pmatrix} 0.4242 & 0.7303 \\ 0.5079 & 0.4886 \\ 0.0855 & 0.5785 \\ 0.2625 & 0.2373 \\ 0.8010 & 0.4588 \\ 0.0292 & 0.9631 \\ 0.9289 & 0.5468 \end{pmatrix}$$

$$+ 2 \begin{pmatrix} 0.2581 & 0.7112 \\ 0.4087 & 0.2217 \\ 0.5949 & 0.1174 \\ 0.2622 & 0.2967 \\ 0.6028 & 0.3188 \end{pmatrix} * \begin{pmatrix} 0.4242 & 0.7303 \\ 0.5079 & 0.4886 \\ 0.0855 & 0.5785 \\ 0.2625 & 0.2373 \\ 0.8010 & 0.4588 \\ 0.0292 & 0.9631 \\ 0.9289 & 0.5468 \end{pmatrix}^\top * \begin{pmatrix} 0.4242 & 0.7303 \\ 0.5079 & 0.4886 \\ 0.0855 & 0.5785 \\ 0.2625 & 0.2373 \\ 0.8010 & 0.4588 \\ 0.0292 & 0.9631 \\ 0.9289 & 0.5468 \end{pmatrix}$$

$$+ 2 * 0.1 \begin{pmatrix} 0.2581 & 0.7112 \\ 0.4087 & 0.2217 \\ 0.5949 & 0.1174 \\ 0.2622 & 0.2967 \\ 0.6028 & 0.3188 \end{pmatrix} = \begin{pmatrix} 0.6866 & -8.9328 \\ -11.4363 & -13.0199 \\ -14.0907 & -12.4113 \\ -14.8288 & -9.4157 \\ -7.3918 & -15.5230 \end{pmatrix}. \tag{4.18}$$

Update $\mathbf{U}$ as:

$$\mathbf{U} = \mathbf{U} - \delta_U * \frac{\partial \mathcal{J}}{\partial \mathbf{U}}$$

$$= \begin{pmatrix} 0.2581 & 0.7112 \\ 0.4087 & 0.2217 \\ 0.5949 & 0.1174 \\ 0.2622 & 0.2967 \\ 0.6028 & 0.3188 \end{pmatrix} - 0.01 * \begin{pmatrix} 0.6866 & -8.9328 \\ -11.4363 & -13.0199 \\ -14.0907 & -12.4113 \\ -14.8288 & -9.4157 \\ -7.3918 & -15.5230 \end{pmatrix} = \begin{pmatrix} 0.2512 & 0.8005 \\ 0.5231 & 0.3519 \\ 0.7358 & 0.2415 \\ 0.4105 & 0.3908 \\ 0.6768 & 0.4740 \end{pmatrix}. \tag{4.19}$$

Similarly, compute $\frac{\partial \mathcal{J}}{\partial L}$,

$$
\frac{\partial \mathcal{J}}{\partial \mathbf{L}} =
\begin{pmatrix}
-4.1231 & -1.6727 \\
-5.9024 & -3.5723 \\
-14.0845 & -12.2855 \\
-2.8054 & -4.1485 \\
-7.4539 & -4.8363 \\
-4.7040 & -7.1421 \\
-13.1224 & -6.2763
\end{pmatrix} ;
\tag{4.20}
$$

update $\mathbf{L} = \mathbf{L} - \delta_L * \frac{\partial \mathcal{J}}{\partial L}$

$$
\mathbf{L} =
\begin{pmatrix}
0.4654 & 0.7471 \\
0.5669 & 0.5243 \\
0.2264 & 0.7014 \\
0.2905 & 0.2788 \\
0.8756 & 0.5072 \\
0.0763 & 1.0345 \\
1.0601 & 0.6096
\end{pmatrix} .
\tag{4.21}
$$

In the second iteration, $\mathbf{U}$ and $\mathbf{L}$ are updated as

$$
\mathbf{U} =
\begin{pmatrix}
0.2455 & 0.8899 \\
0.6386 & 0.4761 \\
0.8821 & 0.3601 \\
0.5641 & 0.4825 \\
0.7649 & 0.6340
\end{pmatrix}
\qquad
\mathbf{L} =
\begin{pmatrix}
0.5040 & 0.7625 \\
0.6241 & 0.5658 \\
0.3734 & 0.8436 \\
0.3167 & 0.3232 \\
0.9631 & 0.5590 \\
0.1156 & 1.1087 \\
1.2007 & 0.6842
\end{pmatrix} .
\tag{4.22}
$$

This process goes iteratively. We output $\mathbf{U}$ and $\mathbf{L}$ after 1000 iterations,

$$
\mathbf{U} =
\begin{pmatrix}
0.0569 & 1.2655 \\
0.4136 & 1.3708 \\
1.6738 & 0.6082 \\
2.6109 & -0.4358 \\
0.5964 & 2.4106
\end{pmatrix}
\qquad
\mathbf{L} =
\begin{pmatrix}
0.4824 & 0.4912 \\
0.0935 & 1.3638 \\
0.8904 & 1.9367 \\
0.0049 & 0.6666 \\
2.6598 & -0.6900 \\
-0.0238 & 1.2734 \\
1.4303 & 1.2145
\end{pmatrix} .
\tag{4.23}
$$

In this example, $\mathbf{U}$ and $\mathbf{L}$ converge after 1000 iterations. Location recommendation can be made based on them. For example, we perform top-1 recommendation to u_5. We multiply $\mathbf{U}$ and $\mathbf{L}$ as the *approximation of check–in preferences to get $\hat{C}$.*

$$\hat{\mathbf{C}} = \mathbf{U} * \mathbf{L}^\top = \begin{pmatrix} 0.6490 & 1.7313 & 2.5016 & 0.8440 & -0.7220 & 1.6101 & 1.6183 \\ 0.8728 & 1.9083 & 3.0232 & 0.9159 & 0.1542 & 1.7357 & 2.2564 \\ 1.1062 & 0.9858 & 2.6681 & 0.4136 & 4.0323 & 0.7345 & 3.1325 \\ 1.0456 & -0.3503 & 1.4808 & -0.2778 & 7.2453 & -0.6171 & 3.2051 \\ 1.4717 & 3.3434 & 5.1997 & 1.6099 & -0.0769 & 3.0553 & 3.7806 \end{pmatrix}.$$

$$(4.24)$$

According to u_5's check–in history in Table 4.1, there are three unvisited locations l_1, l_4, and l_5. The estimated preferences of u_5 on these locations in Eq. (4.24) are 1.4717, 1.6099, and -0.0769, respectively. Thus, l_4 is recommended to u_5.

Another way to solve Eq. (4.9) is based on the closed-form solutions of each variable. By setting the derivation of Eq. (4.15) and Eq. (4.16) to 0, we obtain

$$\mathbf{U}(\mathbf{L}^\top\mathbf{L} + \alpha I) = \mathbf{CL}$$
$$\mathbf{L}(\mathbf{U}^\top\mathbf{U} + \alpha I) = \mathbf{C}^\top\mathbf{U}, \tag{4.25}$$

where $\mathbf{I}$ is an identity matrix. Rewriting the above equations, we obtain the following updating rules:

$$\mathbf{U} = \mathbf{CL}(\mathbf{L}^\top\mathbf{L} + \alpha\mathbf{I})^{-1}$$
$$\mathbf{L} = \mathbf{C}^\top\mathbf{U}(\mathbf{U}^\top\mathbf{U} + \beta\mathbf{I})^{-1} \tag{4.26}$$

which correspond to the closed-form solutions of each variable.

A check-in matrix is usually large-scale and sparse. The computational cost of this algorithm is very high due to the matrix inverse operation for large-scale matrix. Furthermore, taking matrix inverse requires the matrix to be full rank. Since a sparse matrix is commonly rank-deficient, a pseudo-inverse operation with SVD decomposition is required. These constraints make the entire updating process inefficient. Thus, algorithms such as **Gradient Descent** are generally applied for dealing with these issues, while the closed-form solutions is usually applied to small data.

4.2   LOCATION RECOMMENDATION WITH LBSNS

The methods introduced above lay the foundation of location recommendation techniques. However, they do not fully exploit the available information on LBSNs. In Chapter 1, we show the "W⁴" information layout on LBSNs. In this section, we introduce location recommendation approaches specifically designed with these four types of LBSN information, geographical influence, social correlations, temporal patterns, and content indications, as well as their hybrids.

4.2.1   GEOGRAPHICAL INFLUENCE

Geographical influence has significant effects in human mobility. Generally, a user's living area is around some locations, such as home and office, with certain radius [13, 58]. When recommending a location to a user, "how far the location is" is a common concern the user has. Different from sequential-pattern-based approaches which predict visited locations of users, location recommendation with geographical influence focuses on how a user's future check-in on an unvisited location is correlated to his previously visited locations according to their geographical distance.

The "Lévy Flight" property in Section 2.4.2 indicates that the probability of two locations being visited by a user is related to the distance d between them. Denote such probability as $p(d)$. Let $D(l_i, l_j)$ be the geographical distance between l_i and l_j, $I(l_i, l_j)$ be the indication function where $I(l_i, l_j) = 1$ indicates that l_i and l_j were visited by the same user, and 0 otherwise. $p(d)$ can be computed as

$$p(d) = \frac{|\{(l_i, l_j)|D(l_i, l_j) = d, I(l_i, l_j) = 1\}|}{|\{(l_i, l_j)|D(l_i, l_j) = d\}|}, \tag{4.27}$$

which is the number of location pairs with distance d that were visited by the same user, divided by the total number of location pairs with distance d. Figure 4.2 plots $p(d)$ (y-axis) with different d in kilometer (x-axis). The probability presents a power-law like distribution over d, formulated as

$$p(d) \approx a \times d^b, \tag{4.28}$$

where a and b are two parameters that control the power-law distribution.

Location recommendation with geographical influence considers the above power-law distribution [105]. The basic idea is to firstly learn the parameters a and b from the observed check-ins. Then, given an unvisited location l_x of user u, its co-visiting probability is estimated with Equation (4.28) based on its distance to u's previous visited locations.

Denote $\hat{p}(d_{i,j})$ as the estimated probability of l_i and l_j being visited by the same user, approximated with Eq. (4.28). $p(d_{i,j})$ is the observed probability computed with Eq. (4.27). a and b can be learned by minimizing the difference between $\hat{p}(d_{i,j})$ and $p(d_{i,j})$ for all pairs of locations, i.e.,

$$\min_{a,b} \sum_{(l_i, l_j) \in M} (p(d_{i,j}) - \hat{p}(d_{i,j}))^2, \tag{4.29}$$

where M is the set of location pairs in training data.

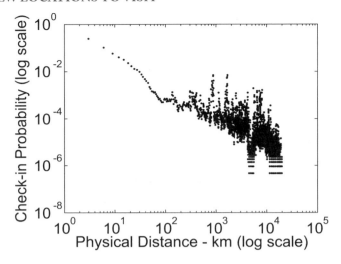

Figure 4.2: Power-law distribution of geographical influence (based on ([105])).

Equation (4.29) can be considered as a least squares problem by applying log function to both $p(d_{i,j})$ and $\hat{p}(d_{i,j})$,

$$\min_{a,b} \sum_{(l_i,l_j)\in M} (\log p(d_{i,j}) - \log \hat{p}(d_{i,j}))^2$$
$$= \sum_{(l_i,l_j)\in M} (\log p(d_{i,j}) - (\log a + b \log d_{i,j}))^2$$

(4.30)

which can be reformulated as

$$\min_{\mathbf{w}} \sum_{(l_i,l_j)\in M} (y_{i,j} - \mathbf{w}^\top \mathbf{x}_{i,j})^2,$$

(4.31)

where

$$y_{i,j} = \log p(d_{i,j}), \mathbf{w} = \begin{pmatrix} \log a \\ b \end{pmatrix}, \mathbf{x}_{i,j} = \begin{pmatrix} 1 \\ \log d_{i,j} \end{pmatrix}.$$

(4.32)

Equation (4.31) is a well-studied least squares problem with closed-form solutions. The regularization term $\frac{1}{2}\|\mathbf{w}\|_F^2$ is also commonly applied to avoid over-fitting. After a and b are learned, one can perform location recommendation. For each unvisited location l_x of u, compute its check-in likelihood $p(l_x)$ as

$$p(l_x) = \prod_{l_j \in C_u} \hat{p}(d_{x,j}),$$

(4.33)

where C_u contains locations visited by u. The location with the maximum $p(l_x)$ will be recommended to u, as it has the highest probability of being visited by u according to its distance to u's previously visited locations.

4.2.2   SOCIAL CORRELATIONS

Social correlations are commonly observed in human mobility especially visiting new locations. For example, a user visits a new restaurant following suggestions from his friends; a user finds an interesting store by asking local friends when coming to a new city. Different from next location prediction, where social information complements spatial and temporal information to generate visited location candidates, location recommendation with social correlations focuses on finding unvisited locations from a user's friends' check-ins based on their social strength.

Location recommendation with social correlations includes "memory-based" and "model-based" approaches, two variants of the corresponding models in general recommender systems.

Memory-Based Social Recommendation

Memory-based social recommendation infers a user's preferences on an unvisited location through his friends' preferences on that location, as shown below:

$$\hat{C_{i,j}} = \bar{C_i} + \frac{\sum_{u_k \in \mathcal{F}(u_i)} \xi(u_i, u_k)(C_{k,j} - \bar{C_k})}{\sum_{u_k \in \mathcal{F}(u_i)} \xi(u_i, u_k)}, \tag{4.34}$$

where $\mathcal{F}(u_i)$ is the set of u_i's friends. Compared to Equation (4.1), a general memory-based approach, it replaces the most similar users with social friends, under the assumption that friends tend to visit similar places due to their social relation. $\xi(u_i, u_k)$ is the tie-strength between u_i and u_k, usually approximated by $sim(u_i, u_j)$ and computed through their check-in history, profile, etc.

Equation (4.1) needs to obtain the similarities between the target user and every other user, while memory-based social recommendation only computes similarities between a target user and his friends. Thus, memory-based social recommendation is more efficient especially for the cases where users join and leave the system frequently. However, there is a trade-off between efficiency and effectiveness. Friends are statically more similar to a user than regular users; while it is not necessary that they are the most similar. Indeed, when investigating the most similar users of a target user, only a small proportion of them are the user's friends. Thus, using only social information may have slightly worse performance than Eq. (4.1).

Model-based Social Recommendation

Model-based social recommendation is an extension of matrix factorization. It applies a social regularization term to Eq. (4.9), as shown below:

$$\min_{\mathbf{U},\mathbf{L}} \|\mathbf{C} - \mathbf{U}\mathbf{L}^\top\|_F^2 + \alpha(\|\mathbf{U}\|_F^2 + \|\mathbf{L}\|_F^2) + \beta Tr(\mathbf{U}^\top \mathcal{L}\mathbf{U}), \tag{4.35}$$

where $Tr(\mathbf{U}^{\top}\mathcal{L}\mathbf{U})$ is commonly referred to as the social regularization term, $\mathcal{L}$ is the Laplacian matrix defined as

$$\mathcal{L} = \mathbf{D} - \mathbf{S}, \tag{4.36}$$

where $\mathbf{S}$ is the tie-strength matrix with $S_{i,j}$ being the tie-strength between u_i and u_j, denoted as $\xi(u_i, u_j)$ shown in Eq. (4.37). $\mathbf{D}$ is a diagonal matrix with $D_{i,i} = \sum_j S_{i,j}$.

$$S = \begin{pmatrix} \xi(u_1, u_1) & \xi(u_1, u_2) & \cdots & \xi(u_1, u_n) \\ \xi(u_2, u_1) & \xi(u_2, u_2) & \cdots & \xi(u_2, u_n) \\ \vdots & \vdots & \ddots & \vdots \\ \xi(u_m, u_1) & \xi(u_m, u_2) & \cdots & \xi(u_m, u_n) \end{pmatrix} \tag{4.37}$$

The social regularization term $Tr(\mathbf{U}^{\top}\mathcal{L}\mathbf{U})$ is derived from

$$\frac{1}{2}\sum_i \sum_j \xi(u_i, u_j)\|\mathbf{U}_i - \mathbf{U}_j\|_2^2. \tag{4.38}$$

Equation (4.38) assumes that a user's interests on latent factors, $\mathbf{U}_i$, are constrained by his tie-strength with his friends. The more similar the two friends are, the closer their latent interests should be. A large $\xi(u_i, u_j)$ would force $\mathbf{U}_i$ to be as close to his friend's latent interests $\mathbf{U}_j$ as possible, while a small $\xi(u_i, u_j)$ could make $\mathbf{U}_i$ loosely fit $\mathbf{U}_j$. The constant $\frac{1}{2}$ is added for calculation convenience. Figure 4.3 illustrates the idea of model-based social recommendation. Equation (4.35) can be solved with the same strategy in Example 4.3, and the recommendation procedure is the same in a general recommender system.

4.2.3　TEMPORAL PATTERNS

Human mobility exhibits strong temporal cyclic patterns. Different from the temporal dynamics approaches for location prediction, location recommendation with temporal patterns focuses on investigating a user's time-dependent check-in interests on a new location. For example, assume that a user has check-ins on l_1, l_2, and l_3 at 13:00, location recommendation aims to infer his interests on a new location (say, l_4) at 13:00, which cannot be solved by the temporal dynamics approaches.

Location recommendation with temporal patterns has a similar idea to collaborative filtering. It infers a user's check-in interests on a new location at a specific time period based on other similar users' check-ins at that time period. However, it needs to address a temporal sparseness problem. For example, among the 24 h of a day, a user does not check-in at each hour; therefore, his check-ins at certain hours could be sparse. This makes it difficult to infer his interests at these hours. Addressing such problems relies on using the temporal properties of temporal cyclic patterns.

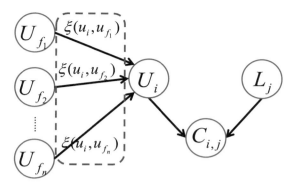

Figure 4.3: Model-based social recommendation.

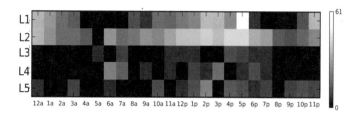

Figure 4.4: Daily check-in activities on LBSNs ([25]).

Two temporal properties are observed in temporal cyclic patterns [25]: (1) **temporal non-uniformness**: a user exhibits distinct check-in preferences at different hours of the day; and (2) **temporal consecutiveness**: a user can have more similar check-in preferences in consecutive hours than in non-consecutive hours. Figure 4.4 plots an illustrative example of a user's aggregated check-ins on his top 5 most visited locations over 24 h on a real-world LBSN datasets. Each cell represents the total number of check-ins at a specific location during the corresponding hour, colored from black (least active) to white (most active). The user's check-in behavior presents a different check-in location distribution at each hour, which changes continually over time.

The two temporal properties can help address the temporal sparseness problem. A user's interests at a specific time period can be inferred not only from other similar users' check-ins at that time period (**temporal non-uniformness**), but also from his own check-ins in nearby time periods (**temporal consecutiveness**). Below we discuss how to model the two temporal properties for location recommendation.

The temporal non-uniformness can be investigated with the following model based on Eq. (4.9),

$$\min_{\mathbf{U}_t,\mathbf{L}} \sum_{t=1}^{T} \|\mathbf{C}_t - \mathbf{U}_t\mathbf{L}^{\top}\|_F^2 + \alpha \sum_{t=1}^{T} \|\mathbf{U}_t\|_F^2 + \beta\|\mathbf{L}\|_F^2. \tag{4.39}$$

Compared to Eq. (4.9), it divides a user's interests on latent factors into T time pieces $\mathbf{U}_t$, with each one inferred from the corresponding check-in matrix $\mathbf{C}_t$. For example, T can be set to 24, with each $\mathbf{C}_t$ consisting of check-ins at a specific hour t of the day, and $\mathbf{U}_t$ being the corresponding user interests at t.

The temporal consecutiveness suggests that a user's preference changes continually over time, which can be modeled as below,

$$\min \sum_{t=1}^{T} \sum_{i=1}^{m} \psi_i(t,t-1)\|\mathbf{U}_t(i,:) - \mathbf{U}_{t-1}(i,:)\|_2^2, \tag{4.40}$$

where $\psi_i(t,t-1) \in [0,1]$ is defined as a temporal coefficient that measures the temporal closeness of u_i's check-in preferences between t and $t-1$. The larger $\psi_i(t,t-1)$ is, the closer u_i's check-in preferences between t and $t-1$. Several similarity measures can be used to compute this value. For example, with cosine similarity, we could obtain

$$\psi_i(t,t-1) = \frac{\mathbf{C}_t(i,:) \cdot \mathbf{C}_{t-1}(i,:)}{\sqrt{\sum_j \mathbf{C}_t^2(i,j)}\sqrt{\sum_j \mathbf{C}_{t-1}^2(i,j)}}, \tag{4.41}$$

when $t = 1$, $U_{t-1} = U_T$. Similar to the social regularization term, Eq. (4.40) also has its matrix form,

$$\min \sum_{t=1}^{T} Tr\left((\mathbf{U}_t - \mathbf{U}_{t-1})^{\top} \Sigma_t (\mathbf{U}_t - \mathbf{U}_{t-1})\right), \tag{4.42}$$

where Σ_t is the diagonal temporal coefficient matrix among m users,

$$\Sigma_t = \begin{bmatrix} \psi_1(t,t-1) & 0 & \cdots & 0 \\ 0 & \psi_2(t,t-1) & \cdots & 0 \\ \vdots & \vdots & \ddots & \vdots \\ 0 & 0 & \cdots & \psi_m(t,t-1) \end{bmatrix}. \tag{4.43}$$

Combining Eq. (4.39) and Eq. (4.42), the final optimization problem with both temporal non-uniformness and temporal consecutiveness properties can be formulated as

$$\min_{\mathbf{U}_t, \mathbf{L}} \sum_{t=1}^{T} \|\mathbf{C}_t - \mathbf{U}_t \mathbf{L}^\top\|_F^2 + \alpha \sum_{t=1}^{T} \|\mathbf{U}_t\|_F^2 + \beta \|\mathbf{L}\|_F^2$$

$$+ \lambda \sum_{t=1}^{T} Tr\big((\mathbf{U}_t - \mathbf{U}_{t-1})^\top \Sigma_t (\mathbf{U}_t - \mathbf{U}_{t-1})\big), \tag{4.44}$$

where λ is a non-negative parameter to control the temporal consecutiveness. It can be solved by the gradient descent algorithm with the similar procedure as shown in Example 4.3. After $\mathbf{U}_t$ and $\mathbf{L}$ are learned, its product $\mathbf{U}_t \mathbf{L}$ can be used to recommend new locations to a user at t.

4.2.4   CONTENT INDICATIONS

Content information on LBSNs could be related to a user's check-ins, providing a unique opportunity for location recommendation from a conceptual perspective. For example, by observing a user's comment on a Mexican restaurant discussing its spicy food, we observe if the user is interested in spicy food or not. This is an example of *user interests*. By observing a location's description as "vegetarian restaurant," we may infer that the restaurant serves "vegetarian food" and users who check-in at this location might be interested in the vegetarian diet. This is an example of *location properties*. These two types of information are representatives of user-generated content and location-associated content on LBSNs. The former refers to comments that are left by users towards specific locations when they check-in; the latter can be descriptive tags associated with specific locations.

Location recommendation with content indications leverages one or two types of the above content information. The fundamental assumption is that users and locations can get connected in the semantic level through geographical topics, as illustrated in Figure 4.5. User-generated content can indicate user's interests on geographical topics, while geographical topics can be assigned to locations by analyzing location-associated content. Note that geographical topics are usually latent topics. Therefore, latent topic models such as LDA or matrix factorization are usually applied to discover them. A matrix factorization model with user-generated content is presented below:

$$\min_{\mathbf{U}, \mathbf{L}, \mathbf{G}} \mathcal{J} = \|\mathbf{C} - \mathbf{U}\mathbf{L}^\top\|_F^2 + \lambda \|\mathbf{A} - \mathbf{U}\mathbf{G}\|_F^2 + \alpha \|\mathbf{U}\|_F^2 + \beta \|\mathbf{L}\|_F^2 + \gamma \|\mathbf{G}\|_F^2, \tag{4.45}$$

where $\mathbf{C}$, $\mathbf{U}$, and $\mathbf{L}$ are check-in matrix, user latent interests, and location latent properties, respectively, as defined before. $\mathbf{A}$ is a user-word matrix extract from user-generated content, and $\mathbf{G}$ represents word distribution over the latent geographical topics. As we can see, the user latent interests in geographical topics $\mathbf{U}$ is factorized from both check-in actions $\mathbf{C}$ and user-generated content **A**. Similarly, location-associated content can be modeled as

$$\min_{\mathbf{U},\mathbf{L},\mathbf{G}} \mathcal{J} = \|\mathbf{C} - \mathbf{U}\mathbf{L}^\top\|_F^2 + \lambda\|\mathbf{B} - \mathbf{L}\mathbf{G}\|_F^2 + \alpha\|\mathbf{U}\|_F^2 + \beta\|\mathbf{L}\|_F^2 + \gamma\|\mathbf{G}\|_F^2, \tag{4.46}$$

where **B** is a location-word matrix extract from location-associated content. Similar to **U**, **L** is factorized from both check-in actions **C** and location-associated content **B**.

The two types of content information can also be used in a unified model,

$$\min_{\mathbf{U},\mathbf{L},\mathbf{G}} \mathcal{J} = \|\mathbf{C} - \mathbf{U}\mathbf{L}^\top\|_F^2 + \lambda_1\|\mathbf{A} - \mathbf{U}\mathbf{G}\|_F^2 + \lambda_2\|\mathbf{B} - \mathbf{L}\mathbf{G}\|_F^2$$
$$+ \alpha\|\mathbf{U}\|_F^2 + \beta\|\mathbf{L}\|_F^2 + \gamma\|\mathbf{G}\|_F^2, \tag{4.47}$$

where user-generated content **A** and location-associated content **B** are factorized into user latent interests **U** and location latent properties **L**, respectively, with a shared word distribution over geographical topics **G**. **U** and **L** together generate check-ins **C**.

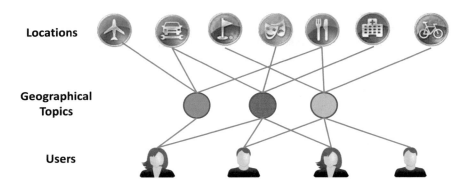

Figure 4.5: Geographical topics connect users and locations (based on ([31])).

4.2.5   HYBRID MODELS

Hybrid models combine different information discussed above for location recommendation. Depending on how multiple types of information are combined, hybrid models can be classified into joint model and fused model.

Joint Model

In a joint model, multiple types of information are considered as a component. Aspects of the component are studied for designing location recommendation models. One typical combination is of geographical and social information, generally referred to as geo-social correlations.

There are strong correlations between geographical distance and social friendship on LB-SNs [17, 83]. The geographical distance plays an important role when constructing social connection between two users, and social connections further affect two user's geographical distance.

Table 4.8: Geo-social groups

	F	$\bar{F}$
$\bar{D}$	$S_{F\bar{D}}$: Local Friends	$S_{\bar{F}\bar{D}}$: Local Non-friends
D	S_{FD} : Distant Friends	$S_{\bar{F}D}$: Distant Non-friends

Therefore, considering both information together could better capture the user preferences for location recommendation on LBSNs.

Location recommendation with geo-social correlations defines four geo-social groups w.r.t. geographical distance D and friendship F, i.e., local friends $S_{F\bar{D}}$, distant friends S_{FD}, local non-friends $S_{\bar{F}\bar{D}}$, and distant non-friends $S_{\bar{F}D}$, as listed in Table 4.8. It assigns any pair of users to one of the geo-social groups.

The geo-social group S_{FD} captures a user's local social correlation, such as going out with friends, following friends' recommendations. $S_{F\bar{D}}$ indicates a user's distant social correlation, such as visiting friends in other states. $S_{\bar{F}D}$ suggests that a user goes to a place where his local neighbors usually go to, which is referred to as confounding effect, i.e., people with the same environment tend to behavior similarly, and visit similar locations. The last geo-social group, $S_{\bar{F}\bar{D}}$, can be explained as an unknown effect, suggesting that a user would randomly visit some new location despite the correlation from his friends or similar users. For example, visiting famous POIs.

Let's define the probability of a user u checking-in at a new location l as $P_u(l)$. With the four geo-social groups, $P_u(l)$ is considered as a weighted combination of the four geo-social correlations, as shown in Eq (4.48). Figure 4.6 illustrates this geo-social recommendation model.

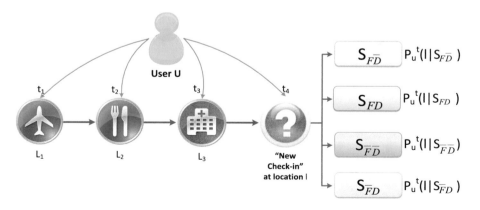

Figure 4.6: The geo-social correlations of check-ins on new locations ([29]).

$$P_u(l) = \phi_1 P_u(l|\mathcal{O}_{S_{\bar{F}\bar{D}}}) + \phi_2 P_u(l|\mathcal{O}_{S_{F\bar{D}}})$$
$$+ \phi_3 P_u(l|\mathcal{O}_{S_{FD}}) + \phi_4 P_u(l|\mathcal{O}_{S_{\bar{F}D}}), \tag{4.48}$$

where ϕ_1, ϕ_2, ϕ_3, and ϕ_4 are parameters that govern the strengths of the four factors, corresponding to geo-social correlation groups, satisfying

$$\phi_1 + \phi_2 + \phi_3 + \phi_4 = 1$$
$$\phi_1 \geq 0, \phi_2 \geq 0, \phi_3 \geq 0, \phi_4 \geq 0. \tag{4.49}$$

$P_u(l|\mathcal{O}_{S_x})$ is the probability of visiting a location l with the correlation from geo-social group $\mathcal{O}_{S_x}$ only. User-based collaborative filtering approaches can be applicable,

$$P_u(l|\mathcal{O}_x) = \bar{C}_u + \frac{\sum_{u_k \in \mathcal{O}_x} sim(u, u_k)(C_{k,l} - \bar{C}_k)}{\sum_{u_k \in \mathcal{O}_x} sim(u, u_k)}. \tag{4.50}$$

The parameters ϕ_1, ϕ_2, ϕ_3, and ϕ_4 can be learned from training data with gradient descent method in Example 4.3. The only difficulty is to meet the constraint in Eq. (4.49). For the non-negativity constraint, one way is to apply a projection strategy which adjusts the value of parameters in each iteration of gradient descent. If ϕ_x in an iteration is negative, it is projected to 0. The projection rule is defined as

$$\begin{cases} 0 & \phi_x < 0 \\ \phi_x & else \end{cases}. \tag{4.51}$$

ϕ_x is then normalized to $[0, 1]$, with $\phi_x = \frac{\phi_x}{\sum_{x=1}^{4} \phi_x}$ to satisfy $\phi_1 + \phi_2 + \phi_3 + \phi_4 = 1$.

In practice, instead of considering ϕ_x as a constant value, it can be defined as a function of features, e.g., a logistic function,

$$\phi_x = f(\mathbf{w}^T \mathbf{f}_u + b)$$
$$= \frac{1}{1 + e^{-(\mathbf{w}^T \mathbf{f}_u + b)}}, \tag{4.52}$$

where $\mathbf{f}_u$ is a feature vector consisting of features related to the user u and corresponding social group $\mathcal{O}_{S_x}$. $\mathbf{w}$ and b are coefficients related to the features. Thus, instead of learning a constant value ϕ_x, it learns feature coefficients $\mathbf{w}$ and b, and makes the model more flexible.

Fused Model
A fused model firstly uses each type of information to obtain recommendation results, and then combines the results together. A fused model usually considers two types of information via the

sum and the product rules. Following the sum rule, results from each type of information are summed up with assigned weights.

$$P_u(l) = \sum_{i=1}^{n} \alpha_i P_u^i(l),$$
(4.53)

where i is the information index, α_i is the weight assigned, and $P_u^i(l)$ is the corresponding recommendation results, indicating the probability of u visiting l based on the analysis of information i.

Applying the product rule, results are multiplied together,

$$P_u(l) = \prod_{i=1}^{n} P_u^i(l).$$
(4.54)

Geographical influence and temporal patterns are combined with either the sum rule [109] or the product rule [17]. Social correlations and geographical influence are fused with the product rule [113]. Content indications and social correlations are fused with the sum rule [101].

Some fused models consider more than two types of information. For example, Ye et al. [105] combined a geographical influence model, a memory-based social recommendation model, and a user-based collaborative filtering model as follows:

$$P_u(l) = \alpha P_u^g(l) + \beta P_u^s(l) + (1 - \alpha - \beta) P_u^{ub}(l),$$
(4.55)

where α and β are parameters to control weights. $P_u^g(l)$ is the probability of u visiting l returned by the geographical influence model, $P_u^s(l)$ is the probability generated by the memory-based social recommendation model, and $P_u^{ub}(l)$ is from the user-based collaborative filtering model.

4.3   EVALUATION METRICS

The performance of location recommendation is evaluated with standard metrics, such as **Precision&Recall** and **RMSE**. Precision evaluates how many locations that are recommended to a user have been visited by the user after recommendation, while recall evaluates how many locations visited by a user have been previously recommended to the user. In the real-world online recommender systems, the number of locations recommended to a user is usually fixed due to the limited slots available on a web page. Thus, top-N recommendation is usually adopted while Precision@N and Recall@N are used. N is the number of recommended locations to a user, usually set to 5 and 10. Precision@N and Recall@N are defined as:

$$precision@N = \frac{\sum_{u_i \in U} |TopN(u_i) \cap L(u_i)|}{\sum_{u_i \in U} |TopN(u_i)|}$$
(4.56)

$$recall@N = \frac{\sum_{u_i \in U} |TopN(u_i) \cap L(u_i)|}{\sum_{u_i \in U} |L(u_i)|},$$
(4.57)

where $TopN(u_i)$ contains top N locations recommended to u_i. $L(u_i)$ contains locations visited by u_i.

As suggested in [103], the effectiveness of recommender systems with sparse datasets (i.e., low-density user-item matrix) is usually not high in terms of Precision@N and Recall@N. For example, the reported Precision@5 is 5% over a dataset with 8.02×10^{-3} density and 3.5% over a dataset with 4.24×10^{-5} density [103, 105]. Thus, algorithms in comparison usually compare the relative performance instead of absolute performance.

The root-mean-square error (**RMSE**) evaluates the difference between predicted values and observed values.

$$RMSE = \sqrt{\frac{\sum_{(i,j)\in C}(\hat{r}_{i,j} - r_{i,j})^2}{|C|}}, \tag{4.58}$$

where $\hat{r}_{i,j}$ represents the predicted value, e.g., estimated check-in count of u_i on l_j, $r_{i,j}$ is the observed value, e.g., observed check-in count. $|C|$ is the size of testing data C.

RMSE is sensitive to the value scale [46]. It is commonly used in product recommendation where the product rating has a fixed range, from 1–5. In location recommendation, the check-in count could vary depending on specific users and locations. Thus, normalization of values in check-in matrix is suggested before applying **RMSE**.

4.4   SUMMARY

Location recommendation, or POI recommendation, has been recognized as an essential task of recommender systems for enriching human life experience. It was first studied on GPS trajectory data. Due to the lack of mapping information between geo coordinates and specific real-world POIs, a POI is usually determined by the stay points (geographical points at which a user spent sufficient long time) extracted from hundreds of users' GPS trajectory logs [116, 118]. With the development of location-based social networking services, users are able to check-in at real-world POIs and share such check-in with their friends through mobile devices, resulting in spatial, temporal, social, and content information to improve location recommendation.

Location recommendation belongs to a sub-category of recommender systems. Thus, techniques of general recommender systems can also be considered for location recommendation with some performance loss. Location recommendation in LBSNs considers the specific human mobility patterns; the corresponding models can be classified into five categories: geographical influence, social correlations, temporal patterns, content indications, and their hybrids. Table 4.9 summarizes the existing work based on the types of LBSN information used.

Although most of the existing work studies more than two types of information, e.g., spatio-temporal, socio-spatial, spatial-content, etc., individual information are commonly combined together through fused method, which restricts the understanding of their deep relationships. In the future, it is possible to study more coherent relationships among multiple types of

Table 4.9: Summary of existing work of location recommendation in LBSNs. The "+" in a cell indicates the corresponding type of information is used in an existing work.

Existing Work	Geographical Influence	Social Correlations	Temporal Patterns	Content Indications
Ye et al. [104], Ye et al. [105], Cheng et al. [13], Gao et al. [29], Zhang et al. [113]	+	+		
Gao et al. [25], Cheng et al. [14]			+	
Ye et al. [103], Long et al. [65], Zhou et al. [119]		+		
Yang et al. [101], Ying et al.[108], Bao et al. [4], Hu et al. [43]		+		+
Yuan et al. [109]	+		+	
Cho et al. [17]	+	+	+	
Gao et al. [28], Gao et al. [26]		+	+	
Liu et al. [63], Yin et al. [106]	+			+
Hu et al. [42], Liu et al. [62], Liu et al. [64], Gao et al. [27]				+
Noulas et al. [72], Chang et al. [11]	+	+	+	+

information, such as the geo-social correlations. This also relies on the discovery of anthropology and social theories of these relationships, which can be helpful for guiding the relationship modeling.

CHAPTER 5

Epilogue

In mining human mobility on location-based social networks, location is an important factor which reflects human interactions with the physical world and is indicative of human activities and lifestyles. As human mobility data surpasses traditional cellphone data in terms of the amount of information it contains, the study of location-based social networks has attracted increasing attentions in recent years. Location-based social networks contain four "W" elements ("who," "when," "where," and "why"), exhibiting various distinct data properties and mobility patterns. It provides a unique opportunity to analyze human mobility from spatial, temporal, social, and content perspectives.

Centered on "location," human mobility on location-based social networks are classified into two typical types of human behaviors: returning to a previous visited location, or going to a new location. We introduced the corresponding algorithms in Chapter 3 and Chapter 4, respectively. We discuss some additional topics related to these two human mobile behaviors.

5.1 LOCATION PRIVACY

Location-based social networks are at the intersection of social media and location-based services. Thus, standard social media mining techniques [110] are applicable to analyzing certain aspects of location-based social networks, such as network measures and online behavior analysis. LBSNs, however, provide a new type of information source - abundant information containing the four "W" elements (when, where, who, and what) regarding a user's real-world and online behavior, and offer an unprecedented opportunity to study human mobility and design advanced location-based services.

Compared with cellphone-based GPS data, LBSNs allow users to have more control of their privacy disclosures, but whenever a user uses LBSN services, he inevitably trades his privacy for convenience. By check-ins, a user can learn about location-specific information including his nearby friends, but the user's location is also shown publicly at the same time. For example, repeated check-ins can reveal one's home or office location [3]. The check-in locations can also indicate personal preference, such as favorite food, regular daily activities, or even health condition [114]. Thus, it results in a trade-off between privacy protection and visibility/convenience.

LBSN services take two strategies to manage the user privacy. The first one comes from the "user-driven" check-in property [24]. Different from GPS data that is passively recorded by mobile devices, a user on LBSNs can choose whether to check-in at a place or not. Thus, if a user has serious privacy concerns when visiting a location, he could choose not to check-in at that location. The second strategy is the privacy setting provided by the service provider or third-party services. It allows a user to select the privacy management options regarding which information he would like to keep private or otherwise.

The above privacy management strategies do not provide a user with the awareness of potential perils on specific check-ins. For example, a user may not consider a big deal to check-in at a restaurant. However, if the restaurant is far away from his home, it could be "valuable" information attracting the attention of burglars.[1] While it is true that the only way to ensure location privacy is not to check-in at any locations, people also deprive of themselves the convenience provided by location-based services. How to achieve the ideal balance between convenience and privacy is beyond the scope of this book.

5.2   LBSN DATA SUFFICIENCY AND RELIABILITY

In the study of human mobility, information loss and information reliability become two major concerns, which correspond to two fundamental issues:

- LBSN data sufficiency, and

- LBSN data reliability.

The check-ins on LBSNs are user-driven: a user usually checks-in at an interested location rather than a regular location such as home or office. Thus, the missing check-in problem can cause information loss, which makes the LBSN data an incomplete representative of the users' mobile traces. However, LBSN data's sufficiency depends on what task to perform. In some tasks, incomplete data could still be sufficient to analyze human mobility. For example, in POI recommendation where new POIs are recommended to a user, the observation of the user's visits at home and office does not add new informative for understanding his check-in preferences at new POIs. While in traffic forecasting, it is suggested to obtain users' mobile traces as complete as possible for predicting traffic.

The check-ins on LBSNs could be "faked" or manipulated by users. When using LBSN services to check-in, a user can manipulate his check-ins by clicking the "check-in" button on a POI without physically being there, such as checking-in at Disneyland while actually sitting at home. There are two major reasons of such "fake" behavior. Some people are for hire to perform fake check-ins and generate fake reviews; and some users may want to earn additional credits through check-ins, which can turn into real-world benefit such as coupons. LBSN services develop strategies to detect such check-ins by verifying a user's actual GPS position w.r.t. his check-in POI at the check-in time. Generally, LBSN data are considered as a reliable information source for mining human mobility, while detecting fake check-ins is still an ongoing challenge.

---

[1]http://www.huffingtonpost.com/2010/02/17/please-rob-me-site-tells_n_465966.html

5.3 CONNECTING "VISITED LOCATIONS" AND "NEW LOCATIONS"

According to the power-law distribution of check-ins, repetitive (i.e., previously *visited*) check-ins and cold-start (or *new*) check-ins fare major forms of human mobility. In Chapter 3, we introduced location prediction to predict whether a user would return to a previously visited location in his next check-in. This is to capture the human repetitive check-ins. Algorithms in this category rely on the analysis of a user's check-in history, including tasks of predicting next visited locations and home locations. In Chapter 4, location recommendation algorithms are discussed which recommend new locations to a user. Subsequently, we introduce spatial, temporal, social and content-based models focusing on new locations.

However, it may not be straightforward to determine which category of algorithms to apply, visited locations or new locations. Most of the time, a user's check-in history is a mixture of both types of locations. One solution is to firstly analyze a user's check-in intention (i.e., determine whether a user would like to go to a visited location or new location in his next check-in), and then apply corresponding prediction or recommendation algorithms. NSM (a Novelty Seeking Model) is such an algorithm for intention analysis [112]. It considers a user's intention of exploring new locations during checking-in as "novelty-seeking trait," and analyze it with a consideration of two aspects: self novelty, i.e., personal desire for exploring, and crowd novelty, i.e., noncompliance with majority behavior.

Another algorithm, CEPR (a Collaborative Exploration and Periodically Returning model) [57] further extends NSM with the consideration of how many opportunities left a user can explore new locations regarding the current status (e.g., current time and location). It applies kernel smoothing techniques on time distribution at a given location to model human mobile regularity, and leverage a Markov model for prediction.

The main drawback of "Novelty Seeking" algorithms comes from its two-step prediction framework. The incorrect prediction of "novelty-seeking trait" in the first step could certainly result in a failure in the second step of prediction/recommendation. It is a still burgeoning area with many novel approaches that are constantly proposed.

5.4 FUTURE DIRECTIONS

Many extensions and work can be further explored. We present some future directions next, hoping to stimulate further discussion and research.

- **Temporal-based Content Analysis**
  Content information has been proven to be useful for mining human mobility patterns. By investigating sentiment and topics embedded in content information, one can infer a user's check-in interests and perform better prediction/recommendation. However, a user's interests may change over the time. A user may not have much seafood before but loves it now due to certain reasons, say, relocation. Such change can be reflected through his check-

in content over a certain period of time. Therefore, temporal-based content analysis could help capture the change of check-in interests and provide the up-to-date indications on human mobility.

- **Anomaly Detection**
  When using location-based social networking services, a user can "pretend" to check-in at one location while physically at another location. As discussed in Section 5.2, users make fake check-ins due to several reasons, such as earning check-in rewards for coupons or free gifts, or being hired to generate positive tips/reviews to gain stellar reputation in a short time. In addition, there are also spam tips containing Phishing links or unsolicited advertisements [1]. Detecting these anomalies not only can improve user's check-in experience, but also can better capture user interests and venue profiles for designing more advanced location-based services.

- **Tensor-Based POI Recommender Systems**
  Temporal information is highly integrated in human mobility, which makes it natural to organize it with other types of information as a tensor. For example, check-in actions can be represented by a tensor with respect to its geographical (longitude and latitude) and temporal dimensions. Thus, tensor-based approaches can be used to study user preferences, which is more compact and intuitive. Furthermore, tensor-based approaches consider different information together, providing an opportunity to study their relationships and complementary effect for personalize mobile applications.

- **Location-Based Mobile Applications**
  Mining human mobility patterns has been an important topic in academic in recent years. As an integral part of human activities, it exhibits great potential for mobile products. Location-based social networking services such as Foursquare and Yelp have already started to analyze users' mobility and perform location recommendation to improve user experience. We expect to see more mobile applications in the next decade which can revolutionarily change ways in facilitating users' daily activities.

Bibliography

[1] A. Aggarwal, J. Almeida, and P. Kumaraguru. Detection of spam tipping behaviour on foursquare. In *Proceedings of the 22nd International Conference on World Wide Web companion*, pages 641–648. International World Wide Web Conferences Steering Committee, 2013. 86

[2] A. Anagnostopoulos, R. Kumar, and M. Mahdian. Influence and correlation in social networks. In *Proceeding of the 14th ACM SIGKDD International Conference on Knowledge Discovery and Data Mining*, pages 7–15. ACM, 2008. DOI: 10.1145/1401890.1401897. 44

[3] L. Backstrom, E. Sun, and C. Marlow. Find me if you can: improving geographical prediction with social and spatial proximity. In *Proceedings of the 19th International Conference on World Wide Web*, pages 61–70. ACM, 2010. DOI: 10.1145/1772690.1772698. 18, 23, 53, 54, 83

[4] J. Bao, Y. Zheng, and M. F. Mokbel. Location-based and preference-aware recommendation using sparse geo-social networking data. In *Proceedings of the 20th International Conference on Advances in Geographic Information Systems*, pages 199–208. ACM, 2012. DOI: 10.1145/2424321.2424348.

[5] S. Barnes and E. Scornavacca. Mobile marketing: the role of permission and acceptance. *International Journal of Mobile Communications*, 2(2):128–139, 2004. DOI: 10.1504/IJMC.2004.004663. 5

[6] H. Bauer, S. Barnes, T. Reichardt, and M. Neumann. Driving consumer acceptance of mobile marketing: a theoretical framework and empirical study. *Journal of Electronic Commerce Research*, 6(3):181–192, 2005. 5

[7] M. Ben-Akiva, M. Bierlaire, H. Koutsopoulos, and R. Mishalani. Dynamit: a simulation-based system for traffic prediction. In *DACCORS Short Term Forecasting Workshop, The Netherlands*. Citeseer, 1998. 5

[8] B. E. Bensoussan and C. S. Fleisher. *Analysis Without Paralysis: 12 Tools to Make Better Strategic Decisions*. FT Press, 2012. 57

[9] F. Cairncross. *The death of distance: How the communications revolution is changing our lives*. Harvard Business Press, 2001. 22

[10] C.-C. Chang and C.-J. Lin.   Libsvm: a library for support vector machines. *ACM Transactions on Intelligent Systems and Technology (TIST)*, 2(3):27, 2011. DOI: 10.1145/1961189.1961199. 48

[11] J. Chang and E. Sun. Location 3: How users share and respond to location-based data on social networking sites. *Proceedings of the Fifth International AAAI Conference on Weblogs and Social Media*, 2011. 18, 49

[12] G. Chen and D. Kotz. A survey of context-aware mobile computing research. *Technical Report TR2000-381, Dept. of Computer Science, Dartmouth College*, 2000. 6

[13] C. Cheng, H. Yang, I. King, and M. Lyu. Fused matrix factorization with geographical and social influence in location-based social networks. *AAAI, Toronto, Canada*, 2012. 69

[14] C. Cheng, H. Yang, M. R. Lyu, and I. King. Where you like to go next: Successive point-of-interest recommendation. In *Proceedings of the Twenty-Third international joint conference on Artificial Intelligence*, pages 2605–2611. AAAI Press, 2013.

[15] Z. Cheng, J. Caverlee, and K. Lee. You are where you tweet: a content-based approach to geo-locating twitter users. In *Proceedings of the 19th ACM International Conference on Information and Knowledge Management*, pages 759–768. ACM, 2010. DOI: 10.1145/1871437.1871535. 51, 52, 54

[16] Z. Cheng, J. Caverlee, K. Lee, and D. Sui. Exploring millions of footprints in location sharing services. In *Proceedings of the Fifth International Conference on Weblogs and Social Media*, 2011. 22, 24, 26, 28, 29, 38

[17] E. Cho, S. Myers, and J. Leskovec. Friendship and mobility: user movement in location-based social networks. In *Proceedings of the 17th ACM SIGKDD International Conference on Knowledge Discovery and Data Mining*, pages 1082–1090. ACM, 2011. DOI: 10.1145/2020408.2020579. 23, 25, 29, 30, 46, 49, 54, 76, 79

[18] S. Consolvo, I. Smith, T. Matthews, A. LaMarca, J. Tabert, and P. Powledge. Location disclosure to social relations: why, when, & what people want to share. In *Proceedings of the SIGCHI Conference on Human Factors in Computing Systems*, pages 81–90. ACM, 2005. DOI: 10.1145/1054972.1054985. 11

[19] J. Cranshaw, R. Schwartz, J. Hong, and N. Sadeh. The livehoods project: Utilizing social media to understand the dynamics of a city. In *Proceedings of the Sixth International AAAI Conference on Weblogs and Social Media*, volume 12, 2012. 9

[20] H. Dia. An object-oriented neural network approach to short-term traffic forecasting. *European Journal of Operational Research*, 131(2):253–261, 2001. DOI: 10.1016/S0377-2217(00)00125-9. 5

[21] N. Eagle, A. Pentland, and D. Lazer. Inferring friendship network structure by using mobile phone data. *Proceedings of the National Academy of Sciences*, 106(36):15274–15278, 2009. DOI: 10.1073/pnas.0900282106. 17, 18

[22] J. Eisenstein, B. O'Connor, N. A. Smith, and E. P. Xing. A latent variable model for geographic lexical variation. In *Proceedings of the 2010 Conference on Empirical Methods in Natural Language Processing*, pages 1277–1287. Association for Computational Linguistics, 2010. 52

[23] H. Gao, G. Barbier, and R. Goolsby. Harnessing the crowdsourcing power of social media for disaster relief. *Intelligent Systems, IEEE*, 26(3):10–14, 2011. DOI: 10.1109/MIS.2011.52. 5

[24] H. Gao and H. Liu. Data analysis on location-based social networks. In *Mobile Social Networking*, pages 165–194. Springer, 2014. DOI: 10.1007/978-1-4614-8579-7_8. 6, 83

[25] H. Gao, J. Tang, X. Hu, and H. Liu. Exploring temporal effects for location recommendation on location-based social networks. In *Proceedings of the 7th ACM conference on Recommender systems*, pages 93–100. ACM, 2013. DOI: 10.1145/2507157.2507182. 73

[26] H. Gao, J. Tang, X. Hu, and H. Liu. Modeling temporal effects of human mobile behavior on location-based social networks. In *Proceedings of the 22nd ACM International Conference on Information & Knowledge Management*, pages 1673–1678. ACM, 2013. DOI: 10.1145/2505515.2505616. 30, 31

[27] H. Gao, J. Tang, X. Hu, and H. Liu. Context-aware point of interest recommendation on location-based social networks. In *Proceedings of the 29th AAAI Conference on Artificial Intelligence*, 2015.

[28] H. Gao, J. Tang, and H. Liu. Exploring social-historical ties on location-based social networks. In *Proceedings of the Sixth International Conference on Weblogs and Social Media*, 2012. 12, 26, 27, 38, 44, 45, 49, 50, 54

[29] H. Gao, J. Tang, and H. Liu. gSCorr: Modeling geo-social correlations for new check-ins on location-based social networks. *21st ACM International Conference on Information and Knowledge Management*, 2012. DOI: 10.1145/2396761.2398477. 8, 12, 77

[30] H. Gao, J. Tang, and H. Liu. Mobile location prediction in spatio-temporal context. *Nokia Mobile Data Challenge Workshop*, 2012. 41, 44

[31] H. Gao, J. Tang, and H. Liu. Personalized location recommendation on location-based social networks. In *Proceedings of the 8th ACM Conference on Recommender Systems*, pages 399–400. ACM, 2014. DOI: 10.1145/2645710.2645776. 5, 7, 21, 76

[32] H. Gao, X. Wang, G. Barbier, and H. Liu. Promoting coordination for disaster relief—from crowdsourcing to coordination. *Social Computing, Behavioral-Cultural Modeling and Prediction*, pages 197–204, 2011. DOI: 10.1007/978-3-642-19656-0_29. 5

[33] M. Gastner and M. Newman. The spatial structure of networks. *The European Physical Journal B-Condensed Matter and Complex Systems*, 49(2):247–252, 2006. DOI: 10.1140/epjb/e2006-00046-8. 22

[34] J. Goldenberg and M. Levy. Distance is not dead: Social interaction and geographical distance in the internet era. *Arxiv preprint arXiv:0906.3202*, 2009. 22

[35] M. C. Gonzalez, C. A. Hidalgo, and A.-L. Barabasi. Understanding individual human mobility patterns. *Nature*, 453(7196):779–782, 2008. DOI: 10.1038/nature06958. 1

[36] M. Goodchild and J. Glennon. Crowdsourcing geographic information for disaster response: a research frontier. *International Journal of Digital Earth*, 3(3):231–241, 2010. DOI: 10.1080/17538941003759255. 5

[37] P. Gundecha, G. Barbier, and H. Liu. Exploiting vulnerability to secure user privacy on a social networking site. In *Proceedings of the 17th ACM SIGKDD Conference*, pages 511–519, 2011. DOI: 10.1145/2020408.2020489. 11

[38] M. Hall, E. Frank, G. Holmes, B. Pfahringer, P. Reutemann, and I. H. Witten. The weka data mining software: an update. *ACM SIGKDD Explorations Newsletter*, 11(1):10–18, 2009. DOI: 10.1145/1656274.1656278. 48

[39] J. Han and M. Kamber. Data mining: concepts and techniques. *San Francisco, CA, itd: Morgan Kaufmann*, 5, 2001. 48

[40] B. Hecht, L. Hong, B. Suh, and E. Chi. Tweets from Justin Bieber's heart: the dynamics of the location field in user profiles. In *Proceedings of the 2011 Annual Conference on Human Factors in Computing Systems*, pages 237–246. ACM, 2011. DOI: 10.1145/1978942.1978976. 51

[41] L. Hong, A. Ahmed, S. Gurumurthy, A. Smola, and K. Tsioutsiouliklis. Discovering geographical topics in the twitter stream. *Proceeding of the 14th ACM SIGKDD International Conference on Knowledge Discovery and Data Mining*, 2012. DOI: 10.1145/2187836.2187940. 9

[42] B. Hu and M. Ester. Spatial topic modeling in online social media for location recommendation. In *Proceedings of the 7th ACM Conference on Recommender Systems*, pages 25–32. ACM, 2013. DOI: 10.1145/2507157.2507174.

[43] B. Hu and M. Ester. Social topic modeling for point-of-interest recommendation in location-based social networks. In *The IEEE International Conference on Data Mining Series*, 2014. DOI: 10.1109/ICDM.2014.124.

[44] X. Hu and H. Liu. Text analytics in social media. *Mining Text Data*, pages 385–414, 2012. DOI: 10.1007/978-1-4614-3223-4_12. 6

[45] L. Humphreys. Mobile social networks and social practice: A case study of dodgeball. *Journal of Computer-Mediated Communication*, 13(1):341–360, 2007. DOI: 10.1111/j.1083-6101.2007.00399.x. 12, 22

[46] R. J. Hyndman and A. B. Koehler. Another look at measures of forecast accuracy. *International Journal of Forecasting*, 22(4):679–688, 2006. DOI: 10.1016/j.ijforecast.2006.03.001. 80

[47] S. W. Jorge Nocedal. *Numerial Optimization*. Springer, 1999. DOI: 10.1007/978-0-387-40065-5. 64

[48] Y. Koren, R. Bell, and C. Volinsky. Matrix factorization techniques for recommender systems. *Computer*, 42(8):30–37, 2009. DOI: 10.1109/MC.2009.263. 57

[49] J. Kulshrestha, F. Kooti, A. Nikravesh, and K. Gummadi. Geographic dissection of the twitter network. *The International Conference on Weblogs and Social Media*, 2012. 23

[50] S. Kumar, F. Morstatter, and H. Liu. *Twitter Data Analytics*. Springer, New York, NY, USA, 2013. DOI: 10.1007/978-1-4614-9372-3. 10

[51] O. Laraki. Twitter places: More context for your tweets. https://blog.twitter.com/2010/twitter-places-more-context-your-tweets, 2010. 4

[52] S. Lederer, J. Mankoff, and A. Dey. Who wants to know what when? privacy preference determinants in ubiquitous computing. In *CHI'03 Extended Abstracts on Human Factors in Computing Systems*, pages 724–725. ACM, 2003. DOI: 10.1145/765891.765952. 11

[53] N. Li and G. Chen. Analysis of a location-based social network. In *International Conference on Computational Science and Engineering*, volume 4, pages 263–270. Ieee, 2009. DOI: 10.1109/CSE.2009.98. 26

[54] Q. Li, Y. Zheng, X. Xie, Y. Chen, W. Liu, and W. Ma. Mining user similarity based on location history. In *Proceedings of the 16th ACM SIGSPATIAL International Conference on Advances in Geographic Information Systems*, page 34. ACM, 2008. DOI: 10.1145/1463434.1463477. 17

[55] D. Lian and X. Xie. Learning location naming from user check-in histories. In *Proceedings of the 19th ACM SIGSPATIAL International Conference on Advances in Geographic Information Systems*, pages 112–121. ACM, 2011. DOI: 10.1145/2093973.2093990. 20

[56] D. Lian and X. Xie.   Mining check-in history for personalized location naming. *ACM Transactions on Intelligent Systems and Technology (TIST)*, 5(2):32, 2014. DOI: 10.1145/2490890. 20

[57] D. LIAN, X. XIE, V. W. ZHENG, N. J. YUAN, F. ZHANG, and E. CHEN. Cepr: A collaborative exploration and periodically returning model for location prediction. *ACM Transactions on Intelligent Systems and Technology*, 2015. 85

[58] D. Lian, C. Zhao, X. Xie, G. Sun, E. Chen, and Y. Rui.   Geomf: joint geographical modeling and matrix factorization for point-of-interest recommendation. In *Proceedings of the 20th ACM SIGKDD International Conference on Knowledge Discovery and Data Mining*, pages 831–840. ACM, 2014. DOI: 10.1145/2623330.2623638. 69

[59] D. Liben-Nowell, J. Novak, R. Kumar, P. Raghavan, and A. Tomkins. Geographic routing in social networks. *Proceedings of the National Academy of Sciences*, 102(33):11623–11628, aug 2005. DOI: 10.1073/pnas.0503018102. 22, 23

[60] J. Lindqvist, J. Cranshaw, J. Wiese, J. Hong, and J. Zimmerman.   I'm the mayor of my house: examining why people use foursquare-a social-driven location sharing application. In *Proceedings of the 2011 Annual Conference on Human Factors in Computing Systems*, pages 2409–2418. ACM, 2011. DOI: 10.1145/1978942.1979295. 12

[61] J. Linshi. Personalizing yelp star ratings: a semantic topic modeling approach. *Yelp Data Challenge*, 2014.

[62] B. Liu, Y. Fu, Z. Yao, and H. Xiong.   Learning geographical preferences for point-of-interest recommendation.   In *Proceedings of the 19th ACM SIGKDD International Conference on Knowledge Discovery and Data Mining*, pages 1043–1051. ACM, 2013. DOI: 10.1145/2487575.2487673.

[63] B. Liu and H. Xiong.   Point-of-interest recommendation in location based social networks with topic and location awareness. *Proc. of SDM'13*, pages 396–404, 2013. DOI: 10.1137/1.9781611972832.44.

[64] X. Liu, Y. Liu, K. Aberer, and C. Miao.   Personalized point-of-interest recommendation by mining users' preference transition. In *Proceedings of the 22nd ACM International Conference on Information & Knowledge Management*, pages 733–738. ACM, 2013. DOI: 10.1145/2505515.2505639.

[65] X. Long and J. Joshi.   A hits-based poi recommendation algorithm for location-based social networks.   In *Proceedings of the 2013 IEEE/ACM International Conference on Advances in Social Networks Analysis and Mining*, pages 642–647. ACM, 2013. DOI: 10.1145/2492517.2492652.

[66] A. Malm. Mobile location-based services. *Berg Insight's LBS Research Series*, 2014. 4

[67] E. Malmi, T. Do, and D. Gatica-Perez. Checking in or checked in: Comparing large-scale manual and automatic location disclosure patterns. *The 11th International Conference on Mobile and Ubiquitous Multimedia (MUM 2012)*, 2012. DOI: 10.1145/2406367.2406400. 26, 38

[68] J. McGee, J. Caverlee, and Z. Cheng. Location prediction in social media based on tie strength. In *Proceedings of the 22nd ACM International Conference on Information & Knowledge Management*, pages 459–468. ACM, 2013. DOI: 10.1145/2505515.2505544. 54, 55

[69] M. McPherson, L. Smith-Lovin, and J. M. Cook. Birds of a feather: Homophily in social networks. *Annual Review of Sociology*, pages 415–444, 2001. DOI: 10.1146/annurev.soc.27.1.415. 7

[70] D. Mok, B. Wellman, and J. Carrasco. Does distance matter in the age of the internet? *Urban Studies*, 47(13):2747, 2010. DOI: 10.1177/0042098010377363. 22

[71] A. Monreale, F. Pinelli, R. Trasarti, and F. Giannotti. Wherenext: a location predictor on trajectory pattern mining. In *Proceedings of the 15th ACM SIGKDD International Conference on Knowledge Discovery and Data Mining*, pages 637–646. ACM, 2009. DOI: 10.1145/1557019.1557091. 12

[72] A. Noulas, S. Scellato, N. Lathia, and C. Mascolo. Mining user mobility features for next place prediction in location-based services. In *ICDM*, pages 1038–1043, 2012. DOI: 10.1109/ICDM.2012.113.

[73] A. Noulas, S. Scellato, C. Mascolo, and M. Pontil. An empirical study of geographic user activity patterns in foursquare. *Proceeding of the 5th International AAAI Conference on Weblogs and Social Media*, 2011. 12, 18, 24, 26

[74] M. Papagelis, D. Plexousakis, and T. Kutsuras. Alleviating the sparsity problem of collaborative filtering using trust inferences. In *Trust Management*, pages 224–239. Springer, 2005. DOI: 10.1007/11429760_16. 63

[75] A. Pozdnoukhov and C. Kaiser. Space-time dynamics of topics in streaming text. In *Proceedings of the 3rd ACM SIGSPATIAL International Workshop on Location-Based Social Networks*, page 8. ACM, 2011. DOI: 10.1145/2063212.2063223. 9

[76] I. Rhee, M. Shin, S. Hong, K. Lee, S. Kim, and S. Chong. On the levy-walk nature of human mobility. *IEEE/ACM Transactions on Networking (TON)*, 19(3):630–643, 2011. DOI: 10.1109/TNET.2011.2120618. 24

[77] N. Sadeh, J. Hong, L. Cranor, I. Fette, P. Kelley, M. Prabaker, and J. Rao. Understanding and capturing people's privacy policies in a mobile social networking application. *Personal and Ubiquitous Computing*, 13(6):401–412, 2009. DOI: 10.1007/s00779-008-0214-3. 12

[78] A. Sadilek, H. Kautz, and J. Bigham. Finding your friends and following them to where you are. In *Proceedings of the Fifth ACM International Conference on Web Search and Data Mining*, pages 723–732. ACM, 2012. DOI: 10.1145/2124295.2124380. 8

[79] A. Sadilek, H. Kautz, and V. Silenzio. Modeling spread of disease from social interactions. In *Sixth AAAI International Conference on Weblogs and Social Media (ICWSM)*, 2012. 9, 11

[80] T. Sakaki, M. Okazaki, and Y. Matsuo. Earthquake shakes twitter users: real-time event detection by social sensors. In *Proceedings of the 19th International Conference on World Wide Web*, pages 851–860. ACM, 2010. DOI: 10.1145/1772690.1772777. 9, 10

[81] S. Scellato, C. Mascolo, M. Musolesi, and V. Latora. Distance matters: Geo-social metrics for online social networks. In *Proceedings of the 3rd Conference on Online Social Networks*, pages 8–8. USENIX Association, 2010. 17, 23

[82] S. Scellato, M. Musolesi, C. Mascolo, V. Latora, and A. Campbell. Nextplace: a spatio-temporal prediction framework for pervasive systems. *Pervasive Computing*, pages 152–169, 2011. DOI: 10.1007/978-3-642-21726-5_10. 12

[83] S. Scellato, A. Noulas, R. Lambiotte, and C. Mascolo. Socio-spatial properties of online location-based social networks. *Proceeding of the 5th International AAAI Conference on Weblogs and Social Media*, 11, 2011. 18, 19, 23, 76

[84] A. Scharl, A. Dickinger, and J. Murphy. Diffusion and success factors of mobile marketing. *Electronic Commerce Research and Applications*, 4(2):159–173, 2005. DOI: 10.1016/j.elerap.2004.10.006. 5

[85] M. Siegler. Yelp enables check-ins on its iphone app; foursquare, gowalla ousted as mayors. *http://techcrunch.com/2010/01/15/yelp-iphone-app-4-check-ins/*, 2010. 4

[86] L. Song, D. Kotz, R. Jain, and X. He. Evaluating location predictors with extensive wi-fi mobility data. In *INFOCOM 2004*, volume 2, pages 1414–1424. IEEE, 2004. DOI: 10.1109/INFCOM.2004.1357026. 34

[87] S. Spaccapietra, C. Parent, M. Damiani, J. De Macedo, F. Porto, and C. Vangenot. A conceptual view on trajectories. *Data and Knowledge Engineering*, 65(1):126–146, 2008. DOI: 10.1016/j.datak.2007.10.008. 12

[88] Stpasha. Normal distribution. *http://en.wikipedia.org/wiki/Normal_distribution*, 2008. 40

[89] X. Su and T. M. Khoshgoftaar. A survey of collaborative filtering techniques. *Advances in Artificial Intelligence*, 2009:4, 2009. DOI: 10.1155/2009/421425. 57

[90] J. Tang, H. Gao, H. Liu, and A. D. Sarma. eTrust: Understanding trust evolution in an online world. In *KDD*, 2012. DOI: 10.1145/2339530.2339574. 57

[91] J. Tang, J. Tang, and H. Liu. Recommendation in social media: recent advances and new frontiers. In *Proceedings of the 20th ACM SIGKDD International Conference on Knowledge Discovery and Data Mining*, pages 1977–1977. ACM, 2014. DOI: 10.1145/2623330.2630813. 7, 57

[92] L. Tang, Y. Jiang, L. Li, and T. Li. Ensemble contextual bandits for personalized recommendation. In *Proceedings of the 8th ACM Conference on Recommender Systems*, pages 73–80. ACM, 2014. DOI: 10.1145/2645710.2645732. 57

[93] Y. Teh. A hierarchical bayesian language model based on pitman-yor processes. In *ACL*, pages 985–992. Association for Computational Linguistics, 2006. DOI: 10.3115/1220175.1220299. 38

[94] N. Thanh and T. Phuong. A gaussian mixture model for mobile location prediction. In *2007 IEEE International Conference on Research, Innovation and Vision for the Future*, pages 152–157. IEEE, 2007. DOI: 10.1109/ICACT.2007.358509. 12

[95] E. Toch, J. Cranshaw, P. Hankes-Drielsma, J. Springfield, P. Kelley, L. Cranor, J. Hong, and N. Sadeh. Locaccino: a privacy-centric location sharing application. In *Proceedings of the 12th ACM International Conference Adjunct papers on Ubiquitous Computing*, pages 381–382. ACM, 2010. DOI: 10.1145/1864431.1864446. 12

[96] J. Tsai, P. Kelley, P. Drielsma, L. Cranor, J. Hong, and N. Sadeh. Who's viewed you?: the impact of feedback in a mobile location-sharing application. In *Proceedings of the 27th International Conference on Human Factors in Computing Systems*, pages 2003–2012. ACM, 2009. DOI: 10.1145/1518701.1519005. 11

[97] D. Wang, D. Pedreschi, C. Song, F. Giannotti, and A. Barabási. Human mobility, social ties, and link prediction. In *Proceedings of the 17th ACM SIGKDD International Conference on Knowledge Discovery and Data Mining*, pages 1100–1108. ACM, 2011. DOI: 10.1145/2020408.2020581. 17

[98] H. Wang, M. Terrovitis, and N. Mamoulis. Location recommendation in location-based social networks using user check-in data. In *Proceedings of the 21st ACM SIGSPATIAL International Conference on Advances in Geographic Information Systems*, pages 374–383. ACM, 2013. DOI: 10.1145/2525314.2525357.

[99] J. Wang, Y. Zhang, C. Posse, and A. Bhasin. Is it time for a career switch? In *Proceedings of the 22nd International Conference on World Wide Web*, pages 1377–1388. International World Wide Web Conferences Steering Committee, 2013. 57

[100] X. Xiao, Y. Zheng, Q. Luo, and X. Xie. Inferring social ties between users with human location history. *Journal of Ambient Intelligence and Humanized Computing*, 5(1):3–19, 2014. DOI: 10.1007/s12652-012-0117-z. 18

[101] D. Yang, D. Zhang, Z. Yu, and Z. Wang. A sentiment-enhanced personalized location recommendation system. *ACM Hypertext*, 2013. DOI: 10.1145/2481492.2481505. 79

[102] M. Ye, K. Janowicz, C. Mülligann, and W. Lee. What you are is when you are: the temporal dimension of feature types in location-based social networks. In *Proceedings of the 19th ACM SIGSPATIAL International Conference on Advances in Geographic Information Systems*, pages 102–111. ACM, 2011. DOI: 10.1145/2093973.2093989. 22, 26, 38

[103] M. Ye, X. Liu, and W. Lee. Exploring social influence for recommendation - a probabilistic generative approach. In *Annual International ACM SIGIR Conference on Research and Development in Information Retrieval*, pages 325–334, 2012. DOI: 10.1145/2348283.2348373. 80

[104] M. Ye, P. Yin, and W. Lee. Location recommendation for location-based social networks. In *Proceedings of the 18th SIGSPATIAL International Conference on Advances in Geographic Information Systems*, pages 458–461, 2010. DOI: 10.1145/1869790.1869861.

[105] M. Ye, P. Yin, W. Lee, and D. Lee. Exploiting geographical influence for collaborative point-of-interest recommendation. In *Annual International ACM SIGIR Conference on Research and Development in Information Retrieval*, pages 325–334, 2011. DOI: 10.1145/2009916.2009962. 5, 57, 69, 70, 79, 80

[106] H. Yin, Y. Sun, B. Cui, Z. Hu, and L. Chen. Lcars: a location-content-aware recommender system. In *Proceedings of the 19th ACM SIGKDD International Conference on Knowledge Discovery and Data Mining*, pages 221–229. ACM, 2013. DOI: 10.1145/2487575.2487608.

[107] Z. Yin, L. Cao, J. Han, C. Zhai, and T. Huang. Geographical topic discovery and comparison. In *Proceedings of the 20th International Conference on World Wide Web*, pages 247–256. ACM, 2011. DOI: 10.1145/1963405.1963443. 9

[108] J. Ying, E. Lu, W. Kuo, and V. Tseng. Urban point-of-interest recommendation by mining user check-in behaviors. In *Proceedings of the ACM SIGKDD International Workshop on Urban Computing*, pages 63–70. ACM, 2012. DOI: 10.1145/2346496.2346507.

[109] Q. Yuan, G. Cong, Z. Ma, A. Sun, and N. M. Thalmann. Time-aware point-of-interest recommendation. In *Proceedings of the 36th International ACM SIGIR Conference on Research and Development in Information Retrieval*, pages 363–372. ACM, 2013. DOI: 10.1145/2484028.2484030. 79

[110] R. Zafarani, M. A. Abbasi, and H. Liu. *Social Media Mining: An Introduction.* Cambridge University Press, 2014. DOI: 10.1017/CBO9781139088510. 83

[111] C. Zhang, L. Shou, K. Chen, G. Chen, and Y. Bei. Evaluating geo-social influence in location-based social networks. In *Proceedings of the 21st ACM International Conference on Information and Knowledge Management*, pages 1442–1451. ACM, 2012. DOI: 10.1145/2396761.2398450. 18

[112] F. Zhang, N. J. Yuan, D. Lian, and X. Xie. Mining novelty-seeking trait across heterogeneous domains. In *Proceedings of the 23rd International Conference on World Wide Web*, pages 373–384. International World Wide Web Conferences Steering Committee, 2014. DOI: 10.1145/2566486.2567976. 85

[113] J.-D. Zhang and C.-Y. Chow. iGSLR: personalized geo-social location recommendation: a kernel density estimation approach. In *Proceedings of the 21st ACM SIGSPATIAL International Conference on Advances in Geographic Information Systems*, pages 324–333. ACM, 2013. DOI: 10.1145/2525314.2525339. 79

[114] X. Zhao, L. Li, and G. Xue. Checking in without worries: Location privacy in location based social networks. In *INFOCOM, 2013 Proceedings IEEE*, pages 3003–3011. IEEE, 2013. DOI: 10.1109/INFCOM.2013.6567112. 83

[115] L. Zheng, C. Shen, L. Tang, T. Li, S. Luis, S.-C. Chen, and V. Hristidis. Using data mining techniques to address critical information exchange needs in disaster affected public-private networks. In *Proceedings of the 16th ACM SIGKDD International Conference on Knowledge Discovery and Data Mining*, pages 125–134. ACM, 2010. DOI: 10.1145/1835804.1835823. 5

[116] V. W. Zheng, Y. Zheng, X. Xie, and Q. Yang. Collaborative location and activity recommendations with gps history data. In *Proceedings of the 19th International Conference on World Wide Web*, pages 1029–1038. ACM, 2010. DOI: 10.1145/1772690.1772795. 80

[117] Y. Zheng, L. Zhang, Z. Ma, X. Xie, and W. Ma. Recommending friends and locations based on individual location history. *ACM Transactions on the Web (TWEB)*, 5(1):5, 2011. DOI: 10.1145/1921591.1921596. 5

[118] Y. Zheng, L. Zhang, X. Xie, and W. Ma. Mining interesting locations and travel sequences from gps trajectories. In *WWW*, pages 791–800. ACM, 2009. DOI: 10.1145/1526709.1526816. 18, 57, 80

[119] D. Zhou, B. Wang, S. Rahimi, and X. Wang. A study of recommending locations on location-based social network by collaborative filtering. *Advances in Artificial Intelligence*, pages 255–266, 2012. DOI: 10.1007/978-3-642-30353-1_22.

[120] K. Zickuhr. Three-quarters of smartphone owners use location-based services. *Pew Internet & American Life Project*, 2012. 4

[121] G. Zipf. Selective studies and the principle of relative frequency in language (cambridge, mass, 1932). *Human Behavior and the Principle of Least-Effort (Cambridge, Mass, 1949*, 1949. 25

Authors' Biographies

HUIJI GAO

Huiji Gao is an applied researcher at LinkedIn. He received his Ph.D. of Computer Science and Engineering at Arizona State University in 2014, and B.S./M.S. from Beijing University of Posts and Telecommunications in 2007 and 2010, respectively. His research interests include social computing, crowdsourcing for disaster management system, recommender systems, and mobile data mining on location-based social networks. He was awarded the 2014 ASU Graduate Education Dissertation Fellowship, the 2014 ASU President's Award for Innovation, the 3rd Place Dedicated Task 2 Next Location Prediction of Nokia Mobile Data Challenge 2012, and Student Travel Awards and Scholarships in various conferences. Updated information can be found at `http://www.nini2yoyo.com`.

HUAN LIU

Huan Liu is a professor of Computer Science and Engineering at Arizona State University. He obtained his Ph.D. in Computer Science at University of Southern California and B.Eng. in Computer Science and Electrical Engineering at Shanghai JiaoTong University. Before he joined ASU, he worked at Telecom Australia Research Labs and was on the faculty at National University of Singapore. He was recognized for excellence in teaching and research in Computer Science and Engineering at Arizona State University. His research interests are in data mining, machine learning, social computing, feature selection, and artificial intelligence, investigating problems that arise in many real-world, data-intensive applications with high-dimensional data of disparate forms such as social media. His well-cited publications include books, book chapters, encyclopedia entries as well as conference and journal papers. He is Editor in Chief of *ACM Transaction on Intelligent Systems and Technology* (TIST), serves on journal editorial boards and numerous conference program committees, and is a founding organizer of the International Conference Series on Social Computing, Behavioral-Cultural Modeling, and Prediction (`http://sbp.asu.edu/`). He is an IEEE Fellow. Updated information can be found at `http://www.public.asu.edu/~huanliu`.